FRED DIBNAH'S BUILDINGS OF BRITAIN

www.rbooks.co.uk

FRED DIBNAH'S BUILDINGS OF BRITAIN

DAVID HALL

BANTAM PRESS

LONDON · NEW YORK · TORONTO · SYDNEY · AUCKLAND

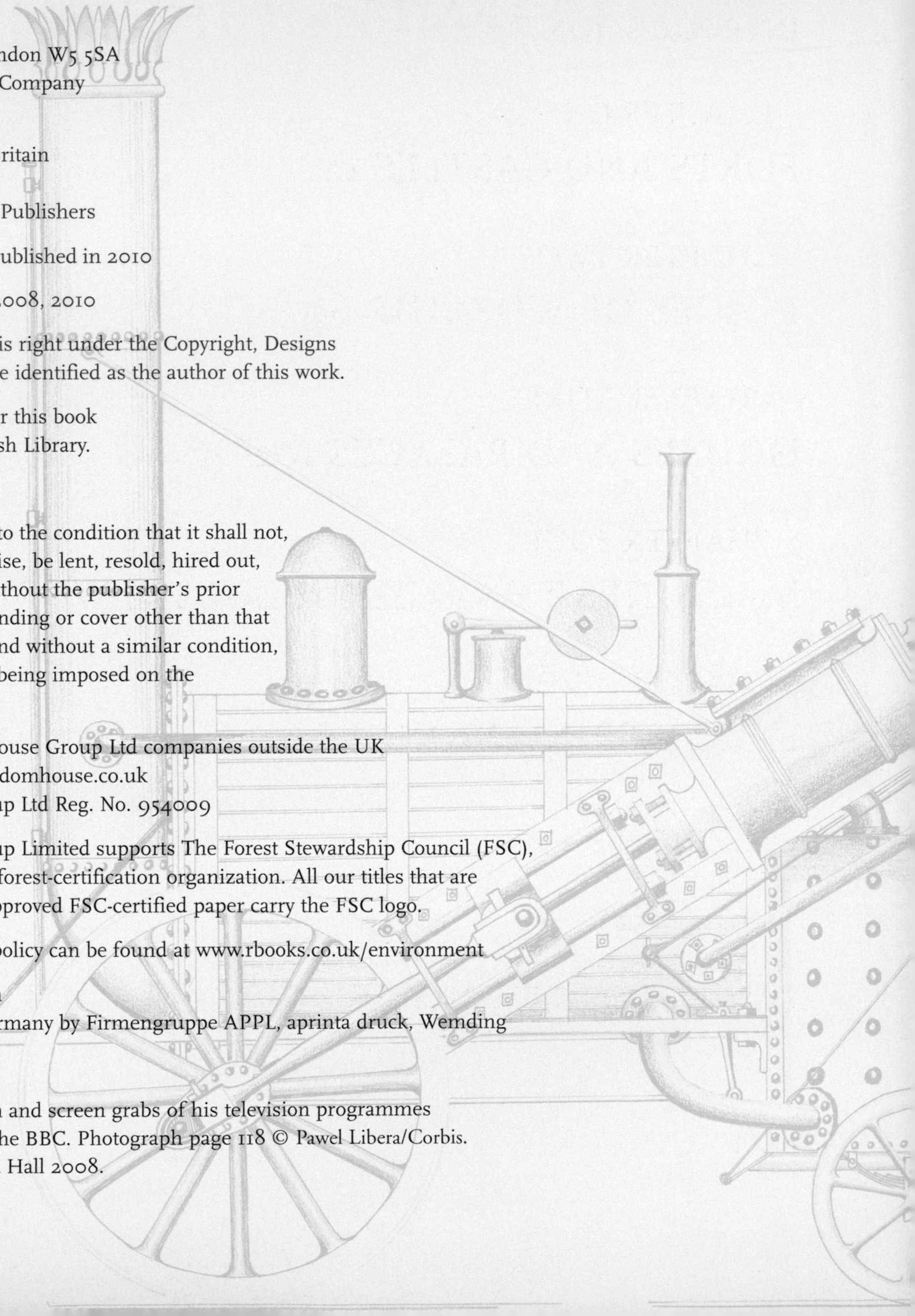

Transworld Publishers
61–63 Uxbridge Road, London W5 5SA
A Random House Group Company
www.rbooks.co.uk

First published in Great Britain
in 2008 by Bantam Press
an imprint of Transworld Publishers

Trade paperback reissue published in 2010

A CIP catalogue record for this book
is available from the British Library.

ISBN 9780593061824

Addresses for Random House Group Ltd companies outside the UK
can be found at: www.randomhouse.co.uk
The Random House Group Ltd Reg. No. 954009

The Random House Group Limited supports The Forest Stewardship Council (FSC), the leading international forest-certification organization. All our titles that are printed on Greenpeace-approved FSC-certified paper carry the FSC logo.

Our paper procurement policy can be found at www.rbooks.co.uk/environment

Designed by Isobel Gillan
Printed and bound in Germany by Firmengruppe APPL, aprinta druck, Wemding

2 4 6 8 10 9 7 5 3 1

Drawings by Fred Dibnah and screen grabs of his television programmes are supplied courtesy of the BBC. Photograph page 118 © Pawel Libera/Corbis.
All other images © David Hall 2008.

CONTENTS

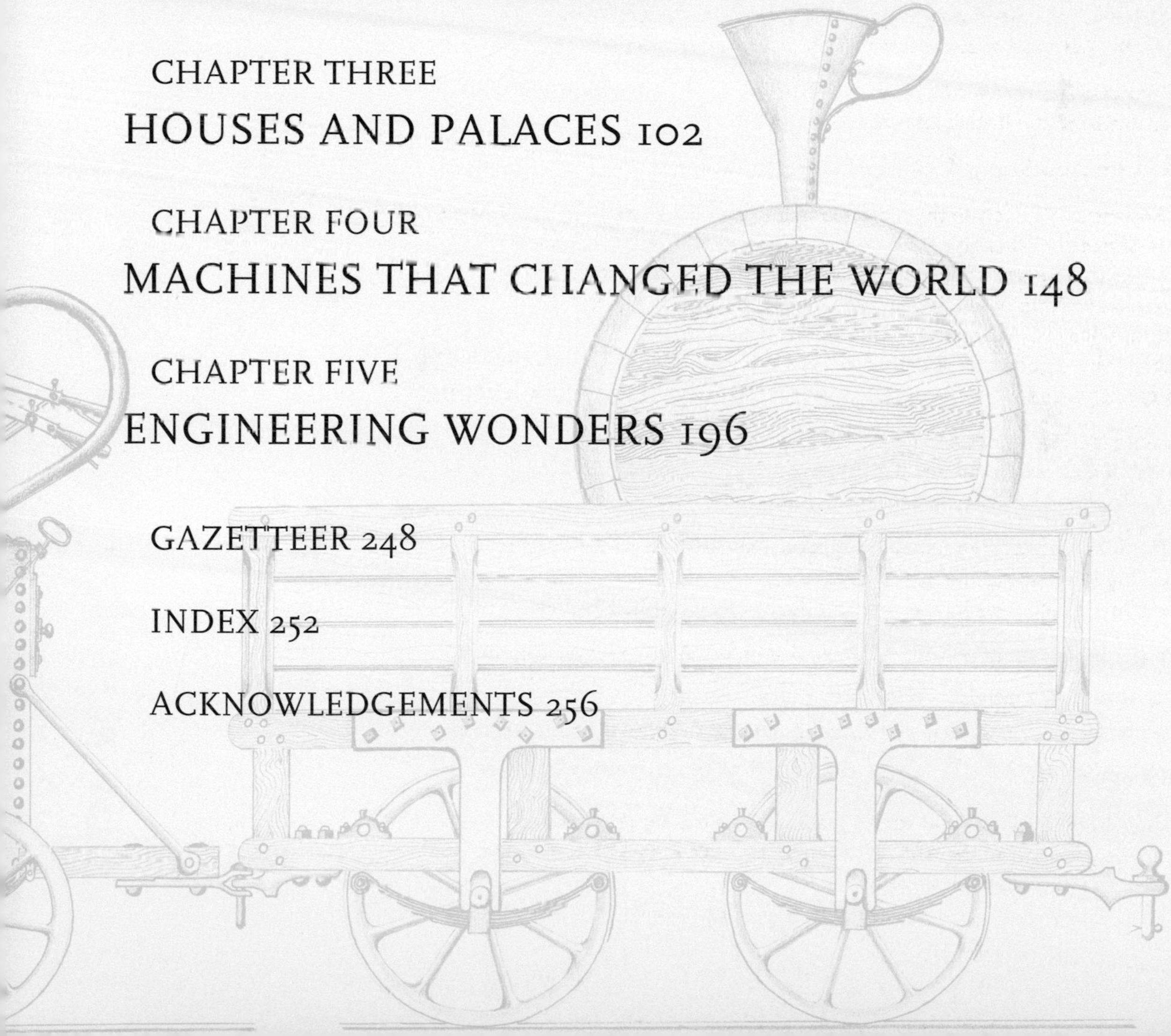

Introduction

Britain is full of magnificent examples of the architectural and engineering genius that we have produced throughout the ages, and of the construction skills of the armies of workers who turned the visions of the architects and engineers into reality. Reminders of our national heritage are all around us. Castles and cathedrals, great houses and palaces, engineering wonders and industrial heritage sites: all have played a part in the building of Britain. For many people, when they visit a great historic building, their first question is: 'How the heck did they build that all that time ago? How did they manage to lift all that stone and wood up to such a great height?'

From tunnels and steam engines to cathedrals: Fred had his own distinctive view of how they were built

It's not just ancient buildings we ask such questions about. What about a marvel of twentieth-century civil engineering like a huge suspension bridge? How on earth do they get the first of those cables suspended between the towers? And how did they build the great steam engines that powered the machines that played their part in the building of Britain when they turned this country into the workshop of the world?

Over a period of seven years from early 1998 until his death in November 2004 I made more than forty programmes with Fred Dibnah for the BBC about our architectural, engineering and

Cragside, home of the Victorian inventor Lord Armstrong, one of Fred's favourite buildings

I've always been quite good at drawing so I can do anything we need to show how things were made

industrial heritage. For seven years we travelled the length and breadth of Britain, visiting magnificent castles and cathedrals, stately homes and palaces; climbing to the top of massive towers and bridges; looking at extraordinary steam-powered pumping engines, mill engines and pit-winding engines; riding on the footplate of famous steam locomotives and walking along the towpaths of the canals that were the arteries of the Industrial Revolution. All of these things are very different in style, in purpose and in appearance but they all had a role in the building of Britain. What fascinated Fred about them more than anything else was the inventiveness of their design and the great range of craft skills of the builders, carpenters, stonemasons, civil and mechanical engineers who built and decorated them. From the remains of Housesteads, the Roman fort we visited on Hadrian's Wall, right up to the great towers of the Humber Suspension Bridge that we climbed to the top of, the skills of the architects and designers and builders are there to be seen.

In all of the programmes we made, what comes across is Fred's distinctive view of these marvels. His interest was not in architectural or engineering theory, but in the practicalities of how things were built. With the help of his beautifully executed drawings he was very good at giving simple, vivid explanations of how it was possible to construct things like great medieval castles and abbeys at a time when technology was limited and there were no power tools, no concrete, no steel, no engines and no heavy machinery. From the huge towers and flying buttresses of our castles and cathedrals to great suspension bridges like Thomas Telford's Conwy Bridge, this book takes us on a tour of some of Britain's most famous historic buildings and gives us Fred's unique perspective on them.

As a young man Fred went to art school and his drawings show him to be a man of considerable artistic talent. And his practical

Engines and Fountains Abbey: what fascinated Fred was the range of skills that went into building them

demonstrations illustrated perfectly how things were built or how things worked, even if some of them did end in disaster. The idea for doing the drawings came when Fred and I were in London researching for the second series we made, *Fred Dibnah's Magnificent Monuments*. While we were at St Paul's Cathedral, the clerk of works, Martin Fletcher, gave Fred a print of a sectional drawing of the dome that illustrated the way that it was constructed. When we got back to our 'digs', as Fred always called the places we stayed, we looked at the drawing and I said it would be good if Fred could use it to show how Christopher Wren had built the dome. 'We don't need to use this,' Fred said. 'I can do one myself. I used to go to art school, you know, and I'm quite good at drawing.' Fred went on to tell me that between leaving school and starting his apprenticeship as a joiner he went to art school in Bolton for a short period:

> I've always been glad I did that. It didn't do me any harm because we used to have forced summer outings with a drawing board and paper and pencil to look at things that were of note in the building line. Learning about perspective and being able to draw a building so that it doesn't look as

Fred's drawing, done to explain the workings of Brunel's atmospheric railway

> though it is going to fall down is very difficult for a lot of people, but I find it quite easy, and I have no problem whatsoever in drawing great big tall factory chimneys that look like great big tall factory chimneys. Not ones like Mr Lowry painted like a black line with a thing across the top and some squiggly lines coming out representing smoke. I've always been quite good at drawing so I can do anything we need to show how things were made.

'Quite good at drawing' turned out to be a bit of an understatement. Fred, it turned out, was a very proficient artist. When we returned from our trip to London, he started work straight away on his own sectional drawing of the dome of St Paul's. After a few days he phoned me to say he'd finished the drawing. When I went to Bolton to have a look I found Fred had done a magnificent job and produced an architectural drawing of the highest order. By this time the programme content and all the locations had been decided on and Fred was keen to do more drawings to help him show how the buildings we were looking at had been constructed, so I selected a number of key architectural features from them and Fred was commissioned to illustrate them. From this stage on and for all the subsequent series, in addition to presenting the

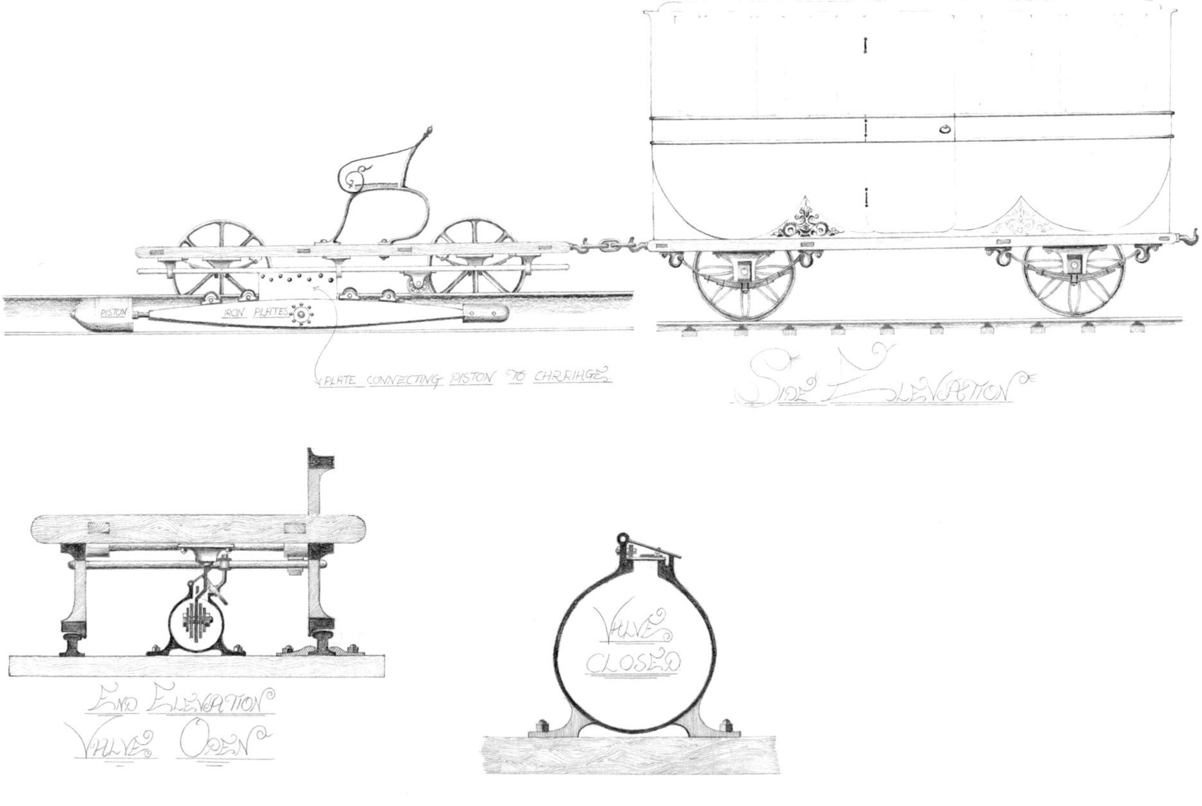

programmes, Fred was also contracted to do a series of drawings. Initially these were to illustrate some of the architectural features of the buildings we were looking at when we were making *Magnificent Monuments* and *Building of Britain*. Later, as we went back to Fred's great passion for steam when we made *Age of Steam*, he turned his artistic skills to some of the most important engines in the history of steam power. Fred's drawings along with his practical demonstrations became a key feature of all of our programmes.

The primary purpose of this book is to provide a showcase for all the beautiful drawings that Fred did for the programmes I made with him and the text has been written around those drawings. It does not, therefore, provide any sort of comprehensive history or overview of British architecture or engineering. Whole architectural ages, styles and movements – Decorated, Perpendicular, Jacobean, Baroque, Palladian – do not get a mention; neither do some of our most illustrious architects like Inigo Jones, Hawksmoor and Vanbrugh because we didn't touch on them in any of our programmes. Likewise 'Machines that Changed the World' ends with Stephenson's *Rocket*, because we didn't use drawings to explain how any engines that followed it worked. The book is, therefore, more of a series of snapshots based around the drawings that Fred did for the programmes. The subject matter ranges over everything we covered, taking in a wide range of architectural styles, engineering skills and different types of construction. I hope, as Fred and I did when we made the programmes, that it will encourage you to visit some of the places that are mentioned and to look at our architectural, engineering and industrial history afresh, with an eye for the skills of the architects, carpenters and stonemasons, builders and construction workers, civil and mechanical engineers who all contributed to the building of Britain.

Ready to drive Tom Brugden's replica of Trevithick's steam carriage

CHAPTER ONE

FORTS AND CASTLES

AFTER FOOD, WATER AND SHELTER, man's most basic need is, arguably, to defend himself. Our earliest defensive constructions were great earthworks like the outer banks and ditches that surround Old Sarum in Wiltshire, about 2 miles north of modern Salisbury. The site was originally a protective hill fort strategically placed at the point where two trade routes met the river Avon. It was constructed during the Iron Age by the local inhabitants who created enormous banks and ditches around the hill. The hill fort is broadly oval in shape, 1,300 feet in length and 1,200 feet in width, and consists of a bank and ditch with an entrance on one of its sides. When we visited, Fred clambered up the outer bank to describe it:

> *When you think this huge earthworks were built about five hundred years BC – it defies all wonder when you think they had no machinery or anything. It was just built by the very basic tools they had and muscle power and when they'd finished all their labours, it must have been pretty effective. For any attacker to get up to the top where the defenders were, they had to climb up this banking with whatever they were going to throw at them, which wouldn't be a lot because they'd be knackered by the time they got up here. Then they'd got to descend into the valley with the wrath of the men up there throwing great rocks at them and then they'd have to attempt to get up out of the ditch on the other side. It must have been a pretty impossible place to take. And when you think all that time ago ... the forerunner of the castle.*

The Romans

Fred at Hadrian's wall; Old Sarum

Hill forts were all pretty basic. For real engineering skills on a truly grand scale you have to wait for the Romans. When the Romans came to Britain they brought with them far more sophisticated building techniques than we'd ever had before and if you want to see the greatest of their works you have to go to the far north of England, close to the Scottish border. Stretching for 73 miles across moors, valleys and crags from Bowness on the Solway Firth in the west to Wallsend-on-Tyne in the east, Hadrian's Wall is the most impressive and lasting legacy of Rome in Britain. It is by far the biggest structure the Romans built here and for around three hundred years it was the north-west frontier of the Roman Empire. The wall was intended to help subdue tribes in the north of Britain by controlling the movement of people in the region, to deter any raids into Roman territory, and to act as a visible symbol of the power of Rome. 'Its purpose,' said Fred, 'was to stop marauding Scotsmen getting across the border, or as Hadrian himself put it "to stop the barbarians getting towards the Romans".' Building work started in AD 122 when Hadrian visited Britain and ordered its construction and it took six years to build. 'They worked bloody hard; it's an amazing piece of work.'

Hadrian's Wall is the most impressive and lasting legacy of the Romans in Britain

Housesteads Fort – North Gate

It was indeed a mammoth task, one of the most ambitious ever undertaken by the Romans, yet it only took six years to build, with at least 8,500 men working in 'centuries': groups of eighty men. The Roman engineers designed it to take advantage of every slope and craggy outcrop of rock along the route, whatever made it harder for an enemy to storm it. The work was carried out by soldiers from the three legions who were stationed in Britain at the time. Within their ranks they had architects, engineers, surveyors, masons and carpenters: everybody they needed for such a huge project.

When it was completed the main garrisons were stationed at sixteen major forts strung out along the wall about 5 miles apart and a line of smaller forts each a Roman mile (1,620 yards) apart. In between each of these milecastles were built two signal towers.

The construction of the wall marked a turning point in Rome's policy towards its far-flung empire. Hadrian, Emperor from AD 117 to 138, was an intelligent and far-sighted leader who could see that the Empire could expand no further. So he decided to create a world of peace and stability within the existing frontiers. In England, behind his protecting wall, he succeeded and Roman civilization flourished for more than 250 years. By around the year AD 400 the Roman Empire had declined in Britain and the wall and its forts were abandoned. 'If it was so well built, though,' said Fred, 'you might wonder why there is so little left of it today. Well, when the Romans left, the wall was abandoned and people began to nick bits of the stone to build their own walls and farmhouses and even churches and abbeys. But even from what is left today you can still see how it was built if you look carefully.' Today it is the most substantial monument of the Roman occupation in Britain and the most remarkable of all Roman frontier works.

Housesteads Fort

Some sections of Hadrian's Wall are much better preserved than others and, 'as you travel along the various parts of the wall today,' Fred observed, 'you can't help but notice how the quality of it all varies as you go from area to area. Some places the stonework just follows the contours of the hillside and other places it's beautiful and level, as though that section had some sort of levelling gear.' All along it there are the remains of forts, turrets and watchtowers. One of the best preserved is the fort at Housesteads, which is the most complete example of a Roman fort in Britain. It stands high on the exposed Whin Sill escarpment with spectacular views. But as well as its commanding position, the site was clearly

chosen because there was a stream to supply constant running water, the location was relatively sheltered, and the land was fertile. The fort was garrisoned by a cohort of around eight hundred infantry, mostly from Belgium, later reinforced by Germanic cavalry. The remains of the governor's house, the very interesting drainage system, barracks, granary, hospital and the toilets or latrines can all be seen on the site.

The walls of the fort were narrow, but backed by a rampart of earth or clay, which sloped up from the structure's interior. The walls were built with sandstone blocks. The turrets on the walls rose to a height of around (30 feet). The main gate at Housesteads was the East Gate, or Porta Praetoria, from which the main street – Via Praetoria – led directly to the headquarters: the Principia. At the various gates of the fort you can see where the iron gudgeons for the hinges of the wooden gates were, and also the grooves in the paving stones caused by the wheels of wagons. At the northern gate Fred looked at what remains.

> This is a rather splendid pillar that once upon a time must have supported two arches in what's left of the northern gateway ... beautiful chisel marks still here. It's amazing after all these years, all these centuries. You can see that there were two towers, one on each side. The far one was once the guardroom I've heard tell. Judging by the thickness of the walls it must have been and it was maybe 30 or 40 feet high. Quite a nice bit of building, no wonder the thing has lasted so long.

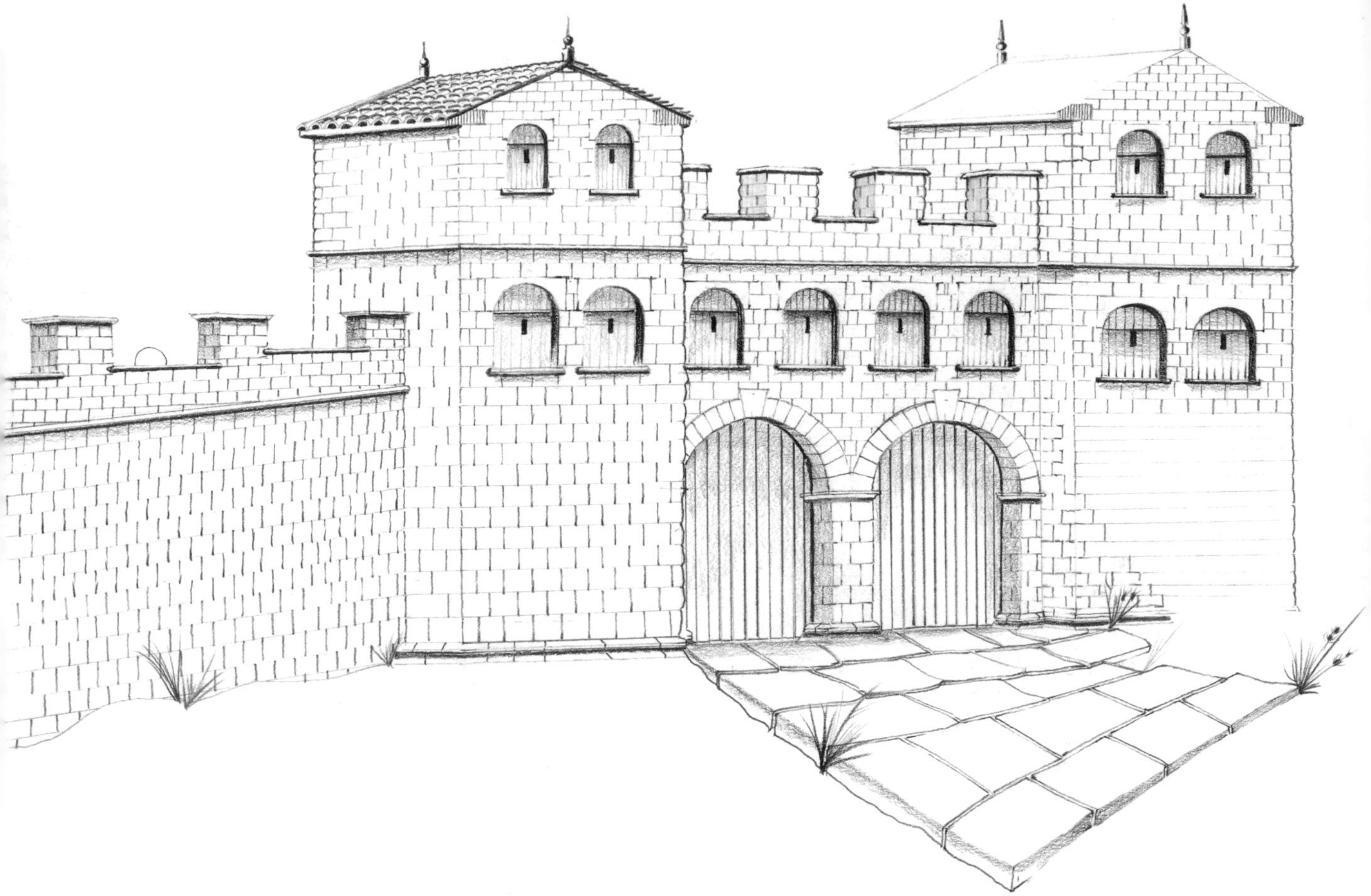

Fred's drawing of this gatehouse was done after we had filmed there to give an impression of what it would have looked like in Roman times.

Inside the fort was an open court, with a colonnade around the south, east and north sides. It had an assembly hall, or basilica: here the orders were issued, and there was a shrine to the imperial cult, where statues of the Emperor were kept. The commanding officer's house was the biggest building on the site and it consisted of a range of rooms around an open courtyard. The actual headquarters building wasn't quite as big; here you can still see how there was an assembly hall in the centre, flanked by a courtyard and by rooms for the administrative staff.

The fort had barracks that could accommodate an infantry cohort of around eight hundred men, but the numbers of men based there varied over time. The barracks were each divided into ten units for the troops, plus larger apartments for the centurion. They had low walls of sandstone, which supported timber frames with wattle-and-daub walls. The fort was garrisoned right up to the end of the Roman occupation

Fred's impression of what the North Gate at Housesteads Fort would have looked like

The latrines at Housesteads – 'It would have been a bit stinky down there in them days'

of Britain, in the early years of the fifth century.

One of the most interesting and best preserved of the remains are the communal latrines in the south-eastern corner of the fort. In here you can see the remains of the very complex arrangements of tanks and channels that ensured the supply of water to the building, and you can see how a main drain carried all the sewage downhill out of the fort. Fred described them:

> This really is one of the highlights of the whole fort: the communal bathtub and communal toilets or latrines. It's rather ingenious how they'd get the water in and you can see how the stonework in the bathtubs were made watertight by beautiful grooves chiselled down in the end of the stones which were then filled in with molten lead. Then they poured molten lead down and then of course they'd caulk it up rather like they'd caulk the planks in the deck of a ship. They didn't waste anything here because the water from the bath came out through an outlet at the end of the bathtub and was channelled into the latrines. It came round a corner and it dripped down into a trough and ran all along the full length of the toilets and back into the main channel that took away all the effluent. The reason for this small trough was that while you were sat on the thunderbox chatting with your mates, you wash your sponge that you used instead of toilet paper in the running water and then after you've done that there were two stone sinks where you washed your hands afterwards. Just what they did with the sponges I don't know, because all the woodwork and the actual planking with the holes in that they sat on are all gone. But I suppose everything ran out down the hill where the sheep are. It would have been a bit stinky down there in them days. But the latrine and the water-supply system that can be seen today and the remains of the hospital at the fort emphasize the Roman army's care for hygiene and for health. The whole thing shows that, as well as being a very ingenious race of people, the Romans were very hygienic as well.

The Normans

Great Hall, Hedingham Castle, Essex; Clifford's Tower, York; Richmond Castle, North Yorkshire

Over six hundred years after the Romans, England was invaded and conquered again, this time by the Normans. The Normans spent the first few decades of their occupation securing the defensive needs of the country and reforming the Saxon church. It was a time when religion went hand in hand with firm military rule, a fact demonstrated by the great number of castles and cathedrals that were built by the Normans in the years immediately after the conquest. These great monuments to Norman power are the earliest buildings of any size that can still be seen in Britain today. Vast stone castles helped to consolidate William's power and great cathedrals and abbeys, commanded by harsh, uncompromising bishops, stood out as citadels of Norman Christianity.

The first castles the Normans built were simple bases for attack and defence in the years immediately after the conquest. These 'ringworks', as they are known, were banks of earth with an outside ditch enclosing a group of wooden buildings. This enclosure, the 'bailey', could be round or oval depending on the arrangement of the buildings and the general lie of the land. William's campaign in 1066–7 involved the building of several of these basic forts in the south-east of England, including the one whose remains can still be

The great castle of Richmond is one of the oldest and finest Norman fortresses in Britain

The keep at Rochester Castle is the largest Norman keep in England

seen within the walls of Dover Castle. Then, as the Normans extended their control over larger parts of England, more of these forts were built so that by the early 1070s there were castles in most parts of England. Quite early it was found that the simple ringwork needed a watchtower. This was raised on a mound or 'motte' inside the bailey. It had its own wooden fence and gate and was reached by a bridge or a gangway up the slope.

During the 1070s the wooden towers and other buildings in the more important forts began to be replaced with stone-built structures. In some the wooden tower standing on top of the motte was replaced by a stone keep. These are known as shell keeps and Clifford's Tower in York is a particularly fine example. In other cases the original motte was abandoned and a separate stone tower was built to replace it, the keep. The stone keep had a strongroom and prison cells in the basement; guardrooms and storerooms on the ground floor; a hall for entertaining and for conducting legal and business affairs on the first floor and the lord's private quarters on the top floor. The Normans built their main permanent castles between the 1070s and 1138, all dominated by massive stone keeps. Colchester, Rochester, Dover, Richmond and the White Tower at the

While the battle raged above, King John's men dug a tunnel to undermine the south tower – a thing that I've done myself many times

Tower of London – situated at major strategic points to guard against invasion or major civil unrest – all date from this period.

Set high on a rocky promontory above the river Swale, the great castle of Richmond is one of the oldest and finest Norman stone fortresses in Britain. Originally built around 1071 by Count Alan Rufus of Richmond, son of the Earl of Brittany, who had fought at the Battle of Hastings, it is one of the few surviving castles with original eleventh-century walls and gatehouse. With a towering keep over 100 feet high added during the reign of Henry II in the 1170s, it retains many of the original features from this period including the chapel and a rare survival of a Norman great hall – 'very probably the oldest in England', according to architectural historian Nikolaus Pevsner.

Many of the greatest castle-builders in the country were bishops, and they helped William the Conqueror stamp his authority on the land with the power of God as well as of the sword. Rochester Castle was built by William de Corbeil, Archbishop of Canterbury from 1123 to 1136, and its mighty keep built of Kentish ragstone is the largest in England, with walls 113 feet high. When we were filming at the castle there was one bit of its history that Fred found particularly fascinating. In 1215 King John laid siege to the castle when it had been garrisoned by rebel barons and the way he did it reminded Fred of the way he did some of his demolition jobs:

> While the battle raged above, King John's men dug a tunnel to undermine the south tower – a thing that I've done myself many times, underpinning a large tower or chimney stack. When I was pulling something down, I used to do it just like this and it was always a bit of a hit-and-miss affair. I nearly always won as King John's men did. The tunnel that they dug is only a few feet below the surface so they had a massive amount of props holding up the fields and the sods and the soil and what have you that was above it. Once they got the tunnel to the base of the tower reputedly they burnt the fat of forty pigs on the pit props to make it burn a bit better and bring its southern corner crashing down. But the keep itself was so strong that it stayed standing. It was taken over by the new King, Henry III, after John's death, who turned it into a royal castle and built the round tower that we see today, so the castle has the original square towers at three of the corners and the new round tower at the fourth. Sadly the keep is an empty shell today. But when it was first built it was a magnificent statement of Norman power.

Fred's drawing of the keep at Hedingham Castle shows how it is made up of arches

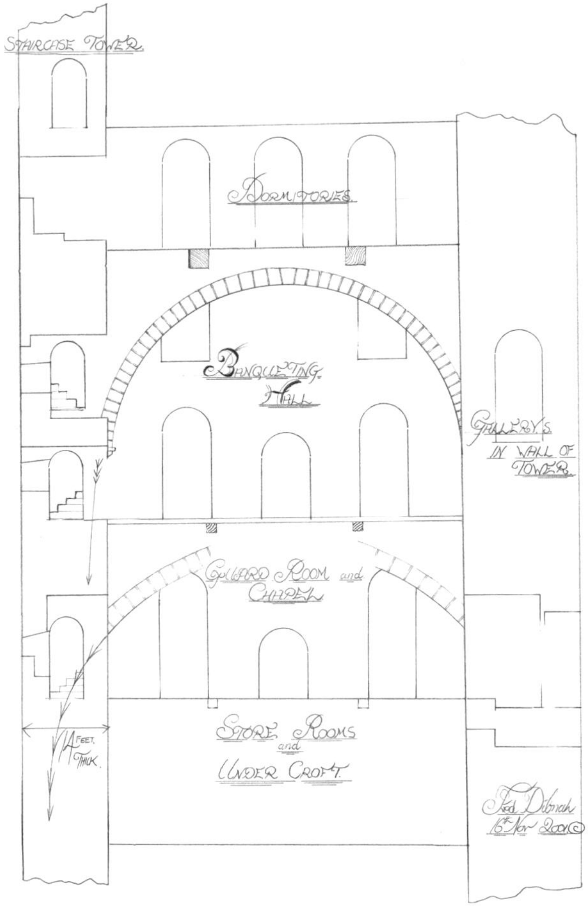

Hedingham Castle

Fred went to Hedingham Castle in Essex to show how the Normans built a keep. Like Rochester it was built by William de Corbeil, Archbishop of Canterbury, and he built it around a huge arch that runs right up its centre, from its foundations to the Great Hall at the top. He did the work for the Norman lord Aubrey de Vere, who was given the lands where the castle stands by William the Conqueror himself. Fred said:

> Aubrey wanted to make his castle look rather posh, so he put an outer skin of dressed stone on it, no doubt to impress his friends and maybe his enemies as well. You can see this outer skin is quite thin really but you get a better idea of what holds the place up downstairs in the undercroft. Down here you can really see what's holding the whole thing up: mortar and lots of it. There's more mortar than there is stone and it's a credit to the men who mixed it because it's still quite solid after all these hundreds of years. You can also see down here, a great pillar that's 14 feet square and goes all the way up to the arch above and it takes the thrust and the whole weight of the building.

Fred's drawing illustrates this.

> This is a cross section of Hedingham Castle keep and as you can see, most of it is made up of arches. But for these arches to stay up they've got to have something pretty big and substantial for them to spring off. Unlike a bridge from one headland to another, you need plenty of meat on each side to take the thrust of the arches. Down in the undercroft, which is the equivalent to the cellar, the walls are actually 14 feet thick with all this great weight of arches pressing

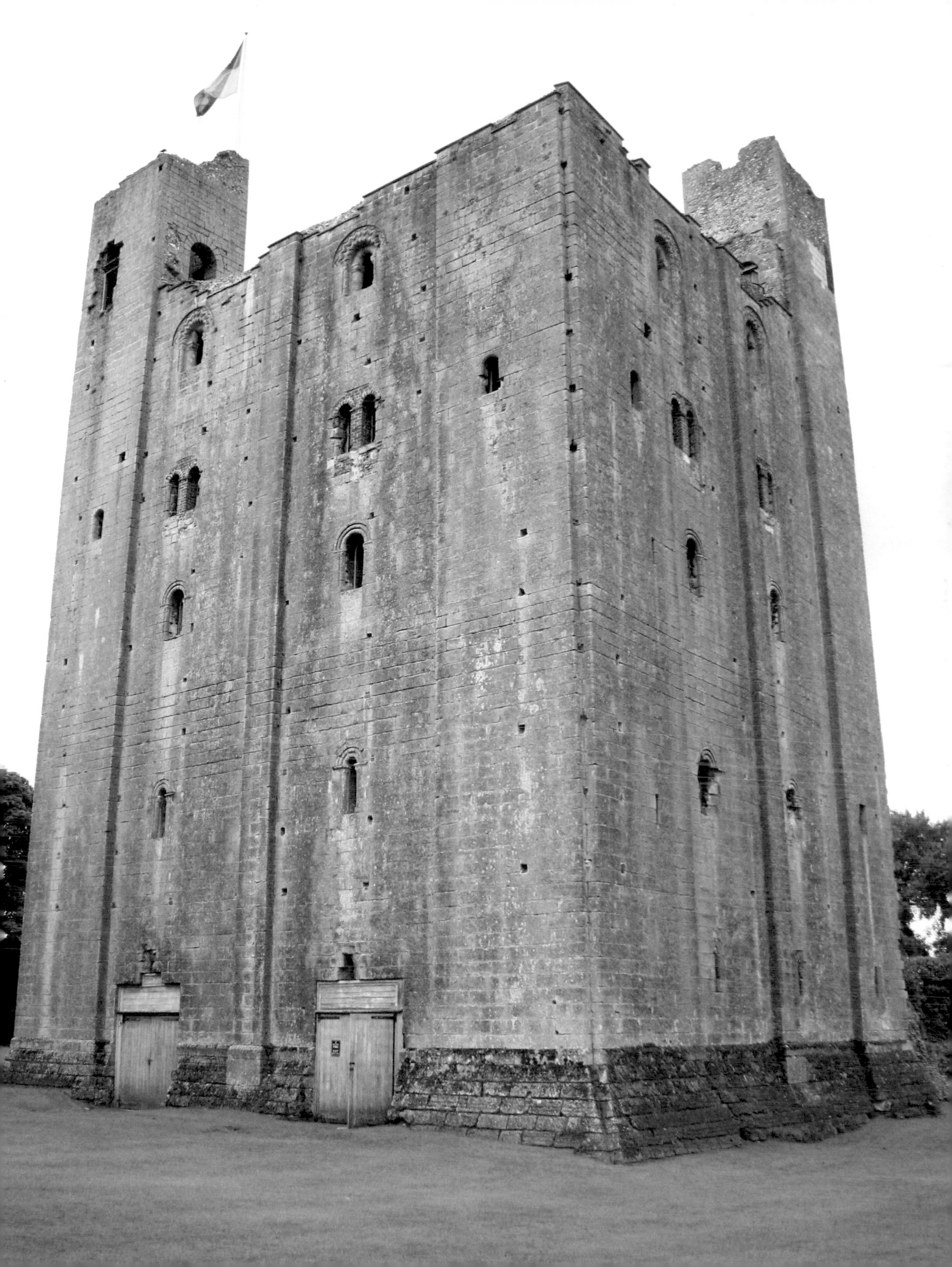

Timber Ceiling
Largest Norman Arch in England
30. Feet. High
28. Feet. Wide

The arch over the banqueting hall at Hedingham Castle

down on them. Then on top of that you've got the weight of a floor, plus I don't know how many knights and noblemen sitting round a great table eating their venison and all that, so you don't want it falling in, do you? Moving further up the building the arch over the banqueting hall is the biggest Norman arch in all of England, something like 28 or 30 feet across.

You can see on the drawing where the great arch emerges out of the walls.

The whole castle really is built on arches. Really, you don't appreciate the arch until you look at the great expanse of floor above, and you can see that it's supporting the great beams that span the ceiling. If they hadn't built the arch of stone they would have had to search around for a tree maybe 50 feet long and 2 feet 6 inches square or something like that and I rather think it were easier to do the arch than find such a tree.

The White Tower at the heart of the Tower of London: an enduring symbol of the Norman Conquest

The White Tower

The castles and cathedrals that the Normans built transformed the face of England and the way the country looked changed just as fundamentally as the way it was ruled. It was with buildings like this that the Norman conquerors established themselves, bringing a stability and permanence to the country that united England under one monarchy. The best-known keep from the Norman period is the White Tower at the heart of the Tower of London. It has stood for over nine hundred years as the enduring symbol of the Norman Conquest. It was built by William to overawe London and put off anybody who might have been thinking about attacking the city. Since then it has been a fortress, a royal residence, a prison, place of execution and a home for the Crown Jewels. Its early history is typical of the great flurry of castle-building activity that came in the years immediately after the Conquest. It goes back to 1066 when William ordered the building of a wooden castle as part of the fortifications to help to secure the chief city of his new kingdom. Ten years later he began the transformation of the fort into a great stone palace with walls 15 feet thick. The work was carried out under the direction of another churchman, Bishop Gundulf of Rochester, who used Kentish ragstone for most of the exterior wall surface with harder limestone brought across the Channel from Caen for dressing buttresses and window surrounds. It became known as the White Tower because, in its early days, the whole building was painted white; today William's great keep still stands proudly right at the centre of the Tower of London. Although heavily restored in the eighteenth century it is still the finest example of a Norman keep in England.

The White Tower was large by the standards of the day because it had to house a strong Norman garrison and occasionally the king and his court. Each floor is divided into three principal apartments. The ground floor was taken up by storerooms, while the king and his court lived on the upper floors when they were in London. Although there have been repairs and alterations to the exterior of the building over the centuries, its shape remains as it was when it was first built.

The Tower remained unchanged for over a century, then in the thirteenth century it was encircled by two towered curtain walls and a great moat. It was Henry III and his son Edward I who carried out these extensions, and by the beginning of the fourteenth century they had turned the original Norman keep into one of the largest strongholds in Europe. The construction of the new outer wall all the way round, with main entrances

from the land by way of a series of gatehouses and drawbridges, transformed the defences of the Tower. Archers and missile-throwing machines along the walls and within the towers that projected from them had a good command of the land around the castle and could concentrate projectiles against an attack at any point. If an enemy did manage to get on to the wall or over it, they were still exposed to missiles from adjoining towers, as well as from the White Tower.

Like the Tower most of our castles have changed a lot over the years. They've had bits added to them: bits pulled down: been changed, modified, turned into places to live, barracks and all manner of things. If you want to see castles that look today very much as they must have done when they were first built, Edward I's Welsh castles are the ones to visit. Of all the great feats of engineering that have helped to shape Britain, there is nothing more dramatic than the great chain of castles Edward built on the coast of North Wales.

The Welsh Castles

Walls and castles at Conwy, Caernarfon and Beaumaris

The Normans had crushed the English within a few years of their invasion, but the Welsh had been very different. This turbulent principality managed to preserve its independence until the reign of Edward I (1272–1307). But when in 1277 the Welsh Prince Llywelyn ap Gruffydd refused to pay Edward homage, the English king vowed to obtain Llywelyn's submission by force of arms. Campaigning in Wales was difficult. The territory was densely wooded, and over two thousand workmen were needed to clear a way through the Welsh forests for Edward's army. Edward fought two hard wars in Wales, and he was determined not to have to fight for Wales ever again. So he began work on medieval Europe's most ambitious building programme. Within the space of twenty-five years he built eight massive stone strongholds which represent the peak of the medieval art of castle-building.

But it wasn't just castles that were built. At Conwy and Caernarfon, the castle was put into a walled town – an idea borrowed from Gascony in southern France, where Edward had been duke. This chain of strongholds was all the work of a Frenchman who revolutionized the art of castle-building: Master James of St George, the greatest military architect of his day. It was while he was working as the Count of Savoy's architect that Master James came up with a whole new style of castle.

Looking down on to Caernarfon Castle from the Eagle Tower

The formidable walls and towers of Harlech Castle stand on a 200-foot rocky spur

For two hundred years, until Master James came up with his revolutionary new designs, the Norman keep like the one at Hedingham had reigned supreme. Fred explained how the new form was developed:

> When Master James came along with his new ideas, it wasn't the actual building techniques that he changed. What he did was much more to do with the overall shape of the castle and its outer defences. Until this time, the keep had formed the heart of the castle. It housed the lord of the manor and it was built on a mound of earth called the motte. Next to this there was an outer stockade where everybody else lived, which was called the bailey. The bailey had a wall round it with a gate in it and of course the gate was the weakest bit of the whole structure. If they could knock the gate down the enemy were in and all the defenders had to run and hide in the keep. What Master James did was to move the keep to the gate and rechristen it the barbican. And what he did next was to do away with the motte all together and build a series of towers round the outer wall. This made the castle much more defendable. If you did manage to breach the walls or gain entry through the barbican, you could be fired at internally by the defenders who would all be on top of the walls and in each of the towers.

Harlech Castle

You can see what Fred was talking about at one of the first of Edward's Welsh castles: Harlech. It's the most defensive of Master James's works, hardly surprising when you think they started building it while Edward was still at war with the Welsh. Fred pointed out that if you look carefully, you can see the different stages of construction marked out on the walls:

> It's very obvious if you look at the main outer wall, you can see at the bottom of it a pretty rough range of stones that look like they were put up in a hurry, to provide a defensive curtain as quickly as possible. It's all quite rough stonework, clearly done by the soldiers while still under attack, not the work of good stonemasons. But that only goes so far up and about 15 feet up from the ground you can see a line in the stone. Above this the stone is much better cut and better dressed, because by now they'd got the time to do it right and at their leisure. Now that they've got a bit of protection from the bottom section of the wall once it had been built, they completed the top 25 or 30 feet in a much better fashion. Better stonemasonry and everything. And then of course, last but not least, the bastion and the curtain wall or outer wall would be built at a later date as an extra form of defence

and if the enemy did approach they could always run inside the castle and leave their trowels and their buckets of mortar for the next time when the enemy had gone away.

Most of the castles were carefully sited so that they could be provisioned from the sea by the English and Harlech was no exception. Seaborne access was vital at times of siege, and although the waters have receded over the centuries, when it was built they lapped right up to the rocks at the foot of the castle crag. It didn't matter if the Welsh were in control of the mountains. It didn't even matter if they were laying siege to the castle, which they did in 1294, because supplies could always be brought up by the 'Way to the Sea', a fortified staircase that plunges 200 feet down the rock to a defended dock below. Men could ferry supplies up these steps completely protected from enemy attack. As you come up the staircase you can see the way the rock had to be dug away to set the foundations of the castle directly on to the solid rock. It's as if the castle was hewn out of the rock itself. The natural strength of the castle rock on the cliff face meant that it was vulnerable from only one side. So they cut a ditch out of the solid rock on the landward side, to protect the castle from attack. They used wedges, and drove them into the cracks in the rock, using the natural plane of the rock to cut it away.

What Master James did was to do away with the motte and build a series of towers round the outer wall!

Conwy Castle

Further up the coast stands Conwy Castle, which was built at the same time as Harlech between 1283 and 1287. The castle and the walled town were complementary and their building proceeded simultaneously. The design and direction of the works was in the hands of James of St George and the master engineer was Richard of Chester. They were in charge of a force of masons, stone-cutters, carpenters and labourers drawn from every corner of England, calculated, when the work was at its height in 1285, to be fifteen hundred strong.

One of the castle's most striking features is the symmetry of its design and the compactness of such a great mass of building. Its distinctive, elongated shape was determined by the narrow rocky outcrop on which it stands. It has eight almost identical towers, four of them on the north and four of them on the south. When you look more closely, though, you can see that the four towers at the eastern end of the castle have an additional turret on the top of them. This is because the castle had two quite separate halves, each one with its own ward or courtyard, and each with its own independent way of entry from the outside. The inner ward was the most secure part of the castle; it could only be approached by water and it was in this part that the king had his accommodation. The four turreted towers are grouped round this ward; they are designed to provide lookout points for watchmen guarding the four corners of the royal residence. The larger outer ward, which could be approached directly from the town, provided accommodation for the castle's permanent garrison. The structure is a piece of monumental engineering on a grand

Conwy Castle's distinctive elongated shape was determined by the narrow rocky outcrop on which it stands

scale, one of the outstanding achievements of medieval military architecture.

The town walls of Conwy are no less impressive. No other town in Britain has such extensive town walls built in one single operation and surviving in an almost completely preserved state. Their purpose was to enclose and protect the town and at the same time to afford, on the landward side, a strong forward defence to the castle itself. The castle and the fortified town were regarded as one unified entity, so much so that day-to-day accounts for the construction work refer to the whole area as the *castrum* instead of distinguishing as they usually did between the *castrum* (castle) and the *villa* (town).

Caernarfon Castle

The grandest of King Edward's castles was Caernarfon, because what he built here was more than just a castle. Caernarfon was sited near the old Roman fort of Segontium and, according to ancient Welsh tradition, was the birthplace of the Roman Emperor Constantine, who had founded the city of Constantinople. So when Edward decided to build Caernarfon, he told Master James to design it in the style of Constantinople with walls of different coloured masonry and polygonal towers. Edward decided that Caernarfon was going to be the centre of his administration in Wales and the castle here would be his royal palace, a symbol of English dominance over the Welsh he had just defeated. The castle would be a showpiece of the new English Empire that had replaced the Empire of Rome.

Work began in July 1283, while Edward was still at war with the Welsh. Houses had to be demolished to make way for part of the castle and it took twenty men five days to cart away their timbers. Then turf and timber palisades were brought in, and a wooden barricade was erected, to defend the building works from enemy attack. A report on the project in 1296 gives an idea of the scale of the operations at Caernarfon. Around 400 skilled masons were in continuous employment on the castle, assisted by nearly 1,000 unskilled labourers who made the mortar and lime, and 30 carpenters. The great stone blocks used for building the castle were all brought to Caernarfon by sea and 30 boats and 160 wagons were needed to keep the supply flowing. Once on site 200 carters were employed to carry the great stones around. This host of craftsmen and labourers on the site were a cosmopolitan crowd who came from nearly every shire in England and from many countries in Europe; so urgent was the work that some masons had to be press-ganged into service.

The plans made provision for a single, massively powerful curtain wall, honeycombed by continuous wall passages at two separate levels to allow the defenders to move along the walls without exposing themselves to enemy fire. Master James built some really clever features into these walls. In one short stretch we can see that there's not enough room to line up enough archers to beat off an attack so it looks as though it could be a weak spot. But what Master James did was to design the arrow slits so that three archers could stand side-by-side and fire across each other simultaneously, trebling the number of archers that could hold this bit of wall.

The circuit of the walls was punctuated by nine towers and two great gatehouses. The

Edward decided that Caernarfon would be his royal palace: a symbol of English dominance over the Welsh

King's Gate was described as the mightiest in the land and to gain entry to the castle here an attacker would have needed to penetrate no fewer than five hefty doors and six portcullises. Traces of their drawbar holes and grooves can still be seen. The Queen's Gate could only be approached by way of a high stone ramp and drawbridge. Most impressive of all is the 128-foot high Eagle Tower, crowned by a triple cluster of turrets, each bearing a stone eagle as a further symbolic link with imperial power. Like the keep of a Norman castle, this resembled a castle in its own right with its own portcullis.

From the top of the Eagle Tower you can see how Master James's defensive scheme works. Caernarfon was defended twice over, by the castle itself and, as at Conwy, by the town walls. Master James was restricted by the shape of his site, but he used its geographical location to the best of his advantage. The castle is built on a peninsula, defended on three sides by water, and the other side is guarded by the castle itself and the town walls. Looking down on to the castle you can see it is shaped like an hourglass, originally divided into two wards by a wall across the middle at the narrowest point.

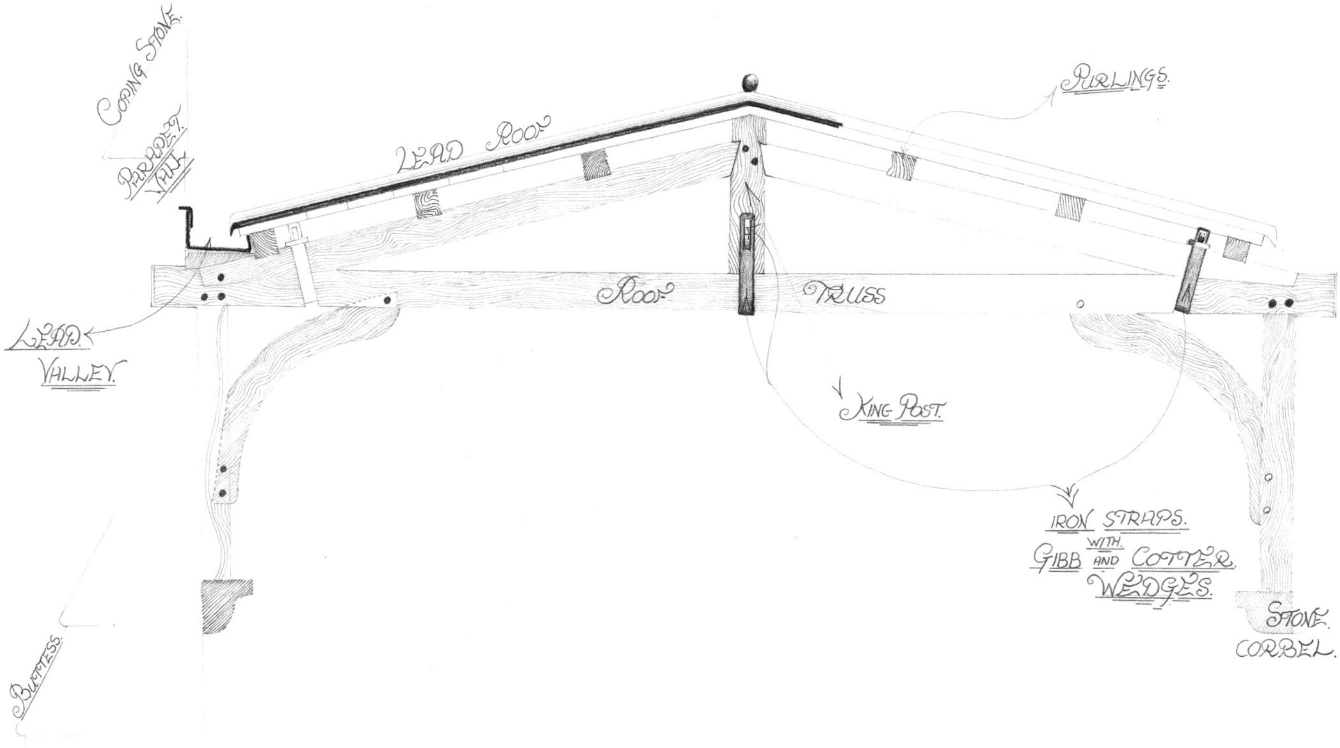

In the lower ward are the remains of the Great Hall and the kitchens. A stretch of curtain wall formed the south side of the Great Hall, 100 feet long. Very little of the hall remains today, but if you know where to look, you can see what it must have been like. Fred's drawing is based on observation of the curtain wall where you can see the remains of the corbelling, which show that the hall had two storeys. If you go inside the Eagle Tower, you will find that here the floors have been reconstructed. The huge beams in the tower are modern ones, but you can see how they sit on the pediment and take the weight of the whole ceiling above them, just as they did in the Great Hall.

Caernarfon was the most expensive of all of Edward's castles. It cost £19,892 9*s* ¾*d* to build, and we know from the exchequer rolls that the total cost of all Edward's castles in Wales was £78,267 11*s* 6*d*. But when he was building these castles, Edward was also at war with Scotland and the combined costs of his wars were making a bit of a dent in the royal finances. There's nowhere better to see this than just across the Menai Strait, on Anglesey. Because here there's another castle which, without any shadow of a doubt, would have been Master James of St George's greatest masterpiece – if only King Edward could have afforded to finish it: Beaumaris.

Fred's drawing of the roof of the Great Hall at Caernarfon based on observation of the curtain wall

Beaumaris Castle

Here Master James had no restrictions on what he could design. It is the most technically perfect castle in the whole of Britain. Like Caernarfon, it has layers of walls within walls, but unlike Caernarfon, it is perfectly symmetrical. And the whole site is surrounded by a moat filled with a controlled supply of tidal water. This was the state of the art for the thirteenth century! The sea has receded here, as at Harlech, but once it came right up to the castle walls and there are no fewer than four successive lines of defence built into this castle. Fred described what an attacker would be confronted with to get into the castle:

> First you had to get across the water-filled moat that surrounded the whole of the castle. Inside the moat is the outer curtain wall with its 8 battlemented sides flanked by 12 battlemented turrets and pierced by 2 twin-towered gateways. An attacker would have to force his way through the gatehouse with boiling oil raining down from the murder holes above and defenders firing at him at point-blank range through arrow slits within the walls that enclosed him. Then, if he survived that, he would be caught within the outer ward, an encircling area of open ground, averaging about 60 feet in width and surrounded by towers and arrow slits, all designed to rain fire down on any attacker who managed to get that far. He'd then be confronted by the massive inner curtain walls, 36 feet high and 15½ feet thick, whose battlements commanded the outer ward. But it's not until you enter the heart of Beaumaris that you get an idea of the sheer scale of its defences. Once you've fought your way inside you still haven't won, because you've got arrows raining down on you from all sides, from the walls. And to get at the king on the far side of the castle you had to fight your way into that building.

The sheer scale of Beaumaris is incredible. Through the gates of its protected harbour, over 2,000 men shifted more than 32,000 tons of stone. They mixed more than 2,000 tons of lime and nailed over 100,000 nails into more than 3,000 boards and that was just in one year!

When you look closely at the walls and the towers, you can get some clues as to how the place was built and with the aid of one of his drawings Fred was able to demonstrate how the medieval masons lifted the stone to build the great walls.

> The builders followed what was a common French practice at the time, which involved using inclined rather than horizontal scaffolds to wheel or drag the heavy loads of

stone to the top as the building gradually rose higher. On the inside of the inner curtain wall you can see a series of three inclined lines of small round holes spaced 6 to 10 inches apart running diagonally up from ground to wall-top level over the whole length of the wall. These are the 'putlog holes', which held the bearer poles of the inclined scaffold paths. As the wall advanced upwards the builders would leave a stone missing and stick a piece of wood into it, then tie a fairpole [main load-bearing vertical pole that supports scaffolding] or a tree trunk to the other end and put boards across which enabled them to raise the materials to the top of the wall as it rose.

They did have cranes but they were too slow so they devised the inclined plane where maybe two or three men could drag a box full of mortar up. Thing is, I still use basically the same methods today. It's much easier to drag a heavy weight than it is to lift it up and carry it. If it's of a reasonable weight and can be dragged along on some sort of sledge it's a lot cheaper and easier than using a crane. It might be a bit slower and more inconvenient but it still works.

Fred's theory was that they formed a continuous loop with the carters going up on one inclined plane and coming down on the other, making the whole building site a bustling hive of activity.

But in spite of this work, Beaumaris was never finished. It was so incredibly expensive that the king simply could not afford to complete it. But, in the work of his builder, James of St George, the medieval art of castle-building had reached its peak in these lonely outposts of English power on the coast of North Wales.

Vast as they were, they were designed so that they could be defended by a small garrison. All the great fortresses of Wales were each manned by fewer than sixty men and they could withstand any siege weapon that medieval man could bring to bear on

Beaumaris – the most technically perfect castle in Britain – has layers of walls within walls

With his drawing, Fred showed how the medieval masons lifted the stone on inclined scaffolds

them – the battering ram, the catapult and the mine. The whole structure was designed to make any attack both difficult and dangerous, if not impossible, and every wall, every tower and every gateway was built with this in mind. Edward died in 1307, and Master James soon followed him, aged over seventy. 'In a way,' said Fred, 'Beaumaris is a monument to their dreams. They both had ideas of grandeur but neither the money nor the time to realize them fully. But you have to admit that together they built something that changed the face of Britain. And in the castles of Edward I, Master James of St George has left us with some of the most impressive structures in the world.'

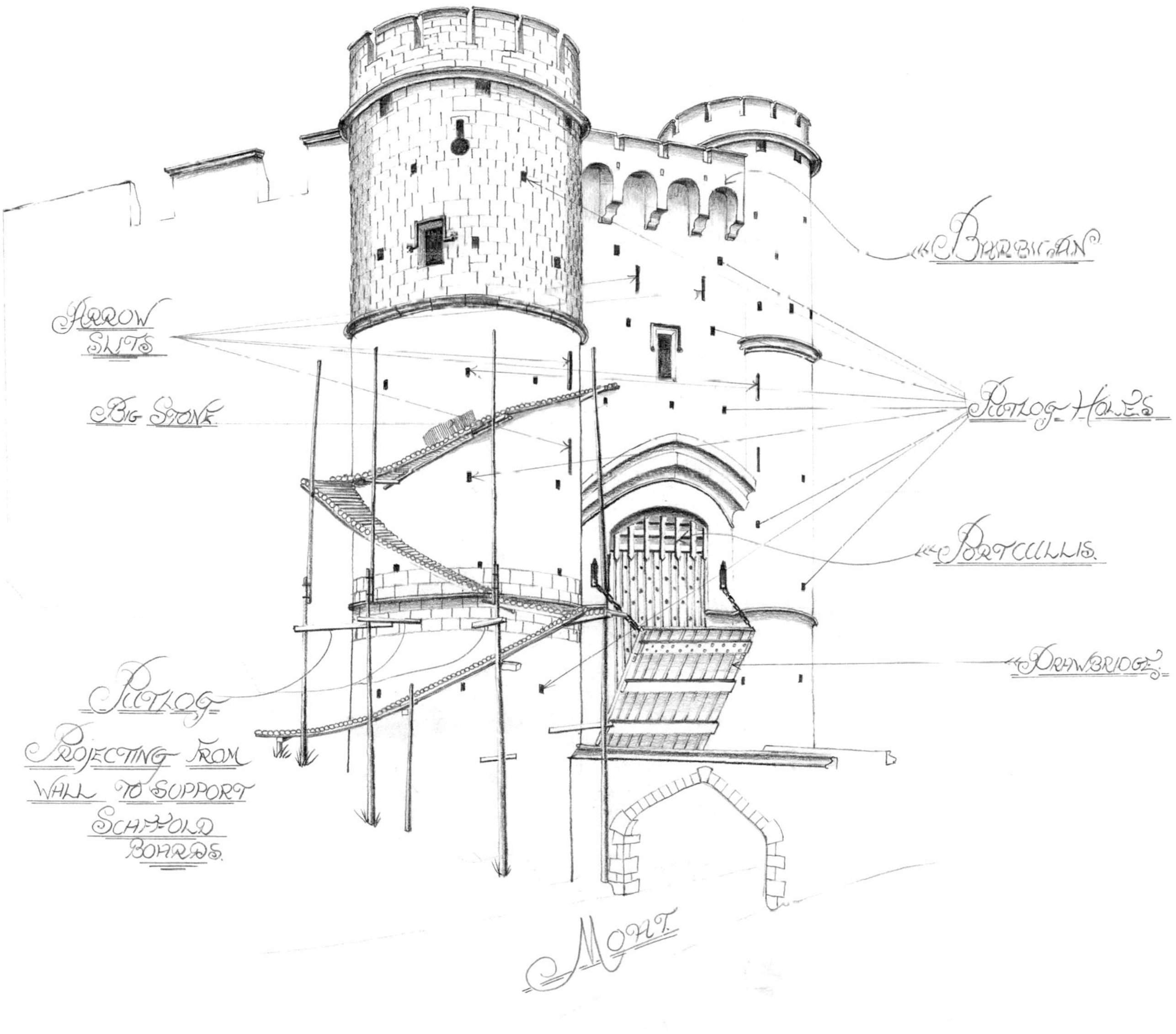

The Late Middle Ages

Today the Welsh castles look very much as they would have done when they were built, but some castles developed in a very different way. Fred went to the heart of England to see Warwick Castle. It's described as one of the finest medieval castles in England, but what he found interesting was the way that it had developed as a place to live. Within the very military-looking defensive walls of the castle there is a magnificent country house. But Warwick's roots go back to the days immediately after the Conquest.

Because the entrance was a favourite target for attackers, it had to be well defended

Warwick Castle

There was a castle here from Norman times and over four centuries Warwick Castle grew from a simple wooden fort built by order of William the Conqueror to the magnificent fortress of the 1480s that we see today. In 1068 the first castle was built on this site – a motte on a cliff overlooking the river Avon with a timber tower standing on top of it. It was part of the chain of command the Normans established to hold recently conquered Saxon England and was fully reconstructed in stone under Henry II. But between 1356 and 1401 the castle began to assume the shape we see today as it underwent extensive reconstruction under the Beauchamp Earls of Warwick.

Caesar's Tower, Warwick Castle

Within the very military looking defensive walls of Warwick Castle, there is a magnificent country house

Work included strengthening the curtain wall against the approach from the town and building two prominent towers at either end. In the middle of the curtain wall there is a massive gatehouse; three storeys high, it is approached through an outer barbican gate and between flanking walls. Fred explained:

> Because the entrance was a favourite target for attackers it had to be fortified as heavily as possible. The barbican had a drawbridge over a wide ditch to give the gatehouse more protection. This was the strongest point of the castle. If you were trying to get in you'd be faced with the first of two iron portcullises and a barrage of crossbow bolts. If for some reason the portcullis didn't get lowered in time, the attackers would find themselves in this narrow roofed passage with arrow slits on either side and, worse, murder holes in the ceiling from which stones, missiles, boiling sand, fire and burning oil would pour down on them. If they managed to survive that there were huge wooden doors to get through. Then they'd have to struggle through more crossfire towards the gatehouse itself. After all that they'd be confronted with another portcullis, another set of murder holes and another door.

The walkways that run along the curtain walls meant that crossbowmen and archers could move quickly to repel attackers at any point around the castle. Once in position, they could pick off the enemy from the battlements. These consisted of solid sections of wall called 'merlots', and gaps that they fired through called 'embrasures'. Towers were the mainstay of the castle's defensive system. Guy's Tower has 12 sides, stands 128 feet high and has 5 storeys. Caesar's Tower has an irregular quatrefoil or cloverleaf shape and rises 147 feet from solid rock just above the river.

Because the towers projected above and out from the wall, they gave archers a clear view downwards and sideways. 'This must have been a pretty impregnable place in the Middle Ages,' Fred declared, 'and even in this modern age, you would have a job to get in there with ordinary firearms and machine-guns if there

were a siege. I wouldn't have liked to try unless there was some really big artillery. But in the Middle Ages they didn't have anything that could do any real damage, other than great slow-moving things that must have weighed at least 7 tons and needed a hundred men to drag them up the road.'

The Middle Ages were pretty violent and turbulent times, but for most castles the last time they saw any real action was at the end of the Civil War in the seventeenth century. After that many of them became palaces and stately homes. Warwick is a good example of the way they changed. From Tudor times onwards the castle took on more of the comforts of a house and by the time James I came to the throne in 1603, the range of domestic buildings had been completed and Warwick Castle was about to become a courtier's palace. The Great Hall is the largest room in the castle and throughout history has been at its heart. It was first constructed in the fourteenth century and rebuilt in the seventeenth. What we see today is the result of restoration work done in the nineteenth century after it had been badly damaged by fire. In the 1890s the castle became a favourite retreat for some of the most important figures in late Victorian society, and the young Winston Churchill was a guest at house parties. Edward, the Prince of Wales, was such a frequent visitor that he had his own bedroom. By this time electricity was the main source of the castle's lighting and there was a steam central heating system. The medieval castle had given way to a gentler way of life as it completed its transformation into a fine stately home.

Castles in Scotland

A similar transition from medieval stronghold to comfortable, aristocratic palace can also be seen in Scotland. From the earliest days there had been the need for defence against the English as well as protection against tribal warfare between clans. At first, as in England, castles took the form of a motte with a timber structure on top and a bailey surrounded by a timber wall. Then stone replaced wood as a more secure form of protection and in the fifteenth and sixteenth centuries the castles became more palatial, displaying their owners' wealth and status.

Fred visited two of Scotland's best-known castles: Edinburgh and Glamis

Edinburgh Castle

Scotland's most famous castle, Edinburgh, has a history that stretches back as far as Warwick and the Tower of London, with the oldest part, St Margaret's Chapel, dating from the Norman period. But most of what we see today evolved later during a stormy history of sieges and wars. Fred described its development:

> Having watched the military tattoo many times on TV, it has never failed to impress me. The whole thing stands on the sheer crag of Castle Rock, the core of an extinct volcano, which rises 435 feet above

Edinburgh Castle stands on the sheer crag of Castle Rock

You can imagine the enemy turning up and saying, 'Sod it, we'll go!'

sea level and it is a pretty formidable natural defence. Building work started off at the top and the castle kept getting a bit bigger as they kept cutting another ledge out of the mountain and building another bit on it. The whole thing is one continuous corkscrew which reminds me of one of those chocolate swirling sweets. You couldn't build a castle on better foundations. It's so high up, literally the top of a volcano. Even now when you're down in town and you look up at all the escarpments that have been chiselled on the edge of the rock as they added to it over the years, it still looks a hell of a long way up. But what it must have looked like when it was first built and there was just the little bit stuck right on top of the rock I don't know. You can imagine the enemy turning up and saying, 'Sod it, we'll go!'

Castle Rock was probably the site of an Iron Age fort which was refortified around the early seventh century. In the eleventh century the castle was the residence of the Scottish king Malcolm III and his pious English queen, Margaret. The Scottish kings continued to use the castle as the royal residence during the following century, but in 1314 all the early buildings except St Margaret's Chapel were destroyed to make the castle useless to the invading English. Building work started again on the rock with the construction of King David's Tower – a large L-shaped keep – in 1367, followed by the King's Lodging in the fifteenth century and the Great Hall in the early sixteenth. The most important part of James IV's (reigned 1473–1513) Great Hall to survive is its hammerbeam roof which looks like an upturned ship's hull. During his reign this was the main banqueting chamber.

In 1650, following Cromwell's invasion of Scotland, the castle was almost completely given over as a garrison and most of its look today reflects its use from this period onwards. Cromwell's Roundhead army, the New Model Army, was a full-time, properly trained and properly paid professional army. A regular army needed accommodation and the castle's Great Hall was converted into a barracks by inserting timber galleries around the walls for soldiers' beds. (These galleries were used until the late 1880s.) After the restoration of the monarchy in 1660 the castle continued to develop as a garrison fortress and new army buildings began to spread out from the upper ward; these included a gunpowder magazine and fireworks factory. The building programme in the castle continued throughout the Victorian age and well into the twentieth century, with new buildings carefully designed to blend in with the existing structure.

Glamis – one of the finest examples of the Scottish baronial style

Glamis Castle

Edinburgh Castle is one of Scotland's best known national monuments, but the most distinctive Scottish achievement was the development of the Scottish Baronial style. This is a style of castle-building adopted by the more prosperous and better educated castle-owners of the late sixteenth and early seventeenth centuries and it is the only thing that can be truly described as a Scottish national style. Its main features are pinnacles and turrets with pointed roofs, corbelled windows and cornices and crow-stepped gables. The lower part of the building was kept bare in the style of the old tower houses from which they were developed and the whole weight of the fanciful decoration was concentrated on the upper part of the building. The tower houses themselves are also a very distinctive Scottish style: simple square towers with the living rooms on the first and upper floors and a vaulted store or stable on the ground floor. They were defensive fortresses built to protect clan chiefs or wealthy noblemen.

Glamis Castle – the childhood home of the late Queen Mother and the legendary setting of *Macbeth* – was described by Sir Walter Scott as the noblest specimen of a medieval castle he had ever seen. It's been a royal residence since the fourteenth century and it's one of the best examples there is of the Scottish Baronial style. The whole thing is a fantastic fairy-tale-castle confection of architectural features.

The earliest known records show Glamis belonging to the Scottish Crown, but the castle as we know it today was begun around 1400 by the second Sir John Lyon when he built what is now the east wing after his marriage to the great-granddaughter of King Robert II. Access to the castle at this time was probably by an external stair to the first floor. His son, Patrick Lyon, was created a peer as Lord Glamis and he commissioned the building of the Great Tower. But this was not linked to the original east wing for another hundred years. The Great Tower, built in the shape of an 'L', was a tall building, typical of many similar castles then being built throughout Scotland. Though primitive by later standards, at the time it would have been fairly impregnable to attack. There would have been a cellar on the ground floor and an entrance hall on the next floor reached by outer stairs. The next floor contained the Great Hall where the lord and his family would live and dine. Surrounding the castle was a fortified court. The 10-foot gap between the Great Tower and the original east wing was filled in with a staircase when the castle was temporarily seized by King James V in the 1530s, but nothing else changed until the remodelling of Glamis in the early seventeenth century by the 1st and 2nd Earls of Kinghorne. It is this that marks its transition from medieval castle to a great house in the Scottish Baronial style, giving the castle its height, its magnificent frontage and its fantastic roof line. And it's this that makes it the finest example of Scotland's most distinctive architectural style. Like Glamis, all of the examples of this style of building are in the north-east of Scotland: great towers which were enlarged, improved upon and magnificently adorned in the Baronial fashion.

At the heart of Glamis Castle you can still see the L-shaped tower house that was built by the first Lord Glamis in the fifteenth century. Because this great sandstone tower was too massive to demolish they simply made a virtue out of necessity by building all around it. The first of the extensions and improvements were done in 1606 when the ninth Lord Glamis was made an earl by King James VI. 'To match his new status,' Fred said, 'he wanted an HQ that looked a bit more impressive than the old tower, so he added a couple of floors to the tower and built a new stair turret. The earl was his own architect and, although there are no records, it is believed that he employed masons of the Aberdeen school under the direction of John Bell, and what a magnificent job they did.'

The alterations were begun in the first decade of the seventeenth century by the first earl and were not finished until the early 1620s, by which time he had been succeeded by his son John, the second earl.

At the heart of Glamis you can see the great fifteenth-century tower house

They added two new floors of chambers and galleries and an attic to the top of the Great Tower. It is this that gives the building its great height and its fantastic roof-line. They also built the great circular stair turret at the front in the angle of the Great Tower to give access to the new upper floors; reworked much of the stonework; provided the elaborate frontage with the corbelled-out turrets and pointed roofs and enriched the interior with plaster ceilings and fireplaces in the latest fashions. 'His builders made a pretty good job of it,' Fred observed, 'because at the front you can't really see any line where they built on top. It's only round the back where you can see any traces of it.'

The new staircase that was added by the first earl is particularly impressive. It was built on the grandest scale, winding round a hollow central pier and it's well lit by windows all the way up. The original reason for this central pier being hollow is not certain, but at one time it was used for some form of central heating as you can see from the brass ducts in it. It's now used for housing the weights for the clock at the top of the tower. Externally the stair tower is richly ornamented with two tiers of armorial panels.

With the aid of his drawing, Fred explained how the turrets were built

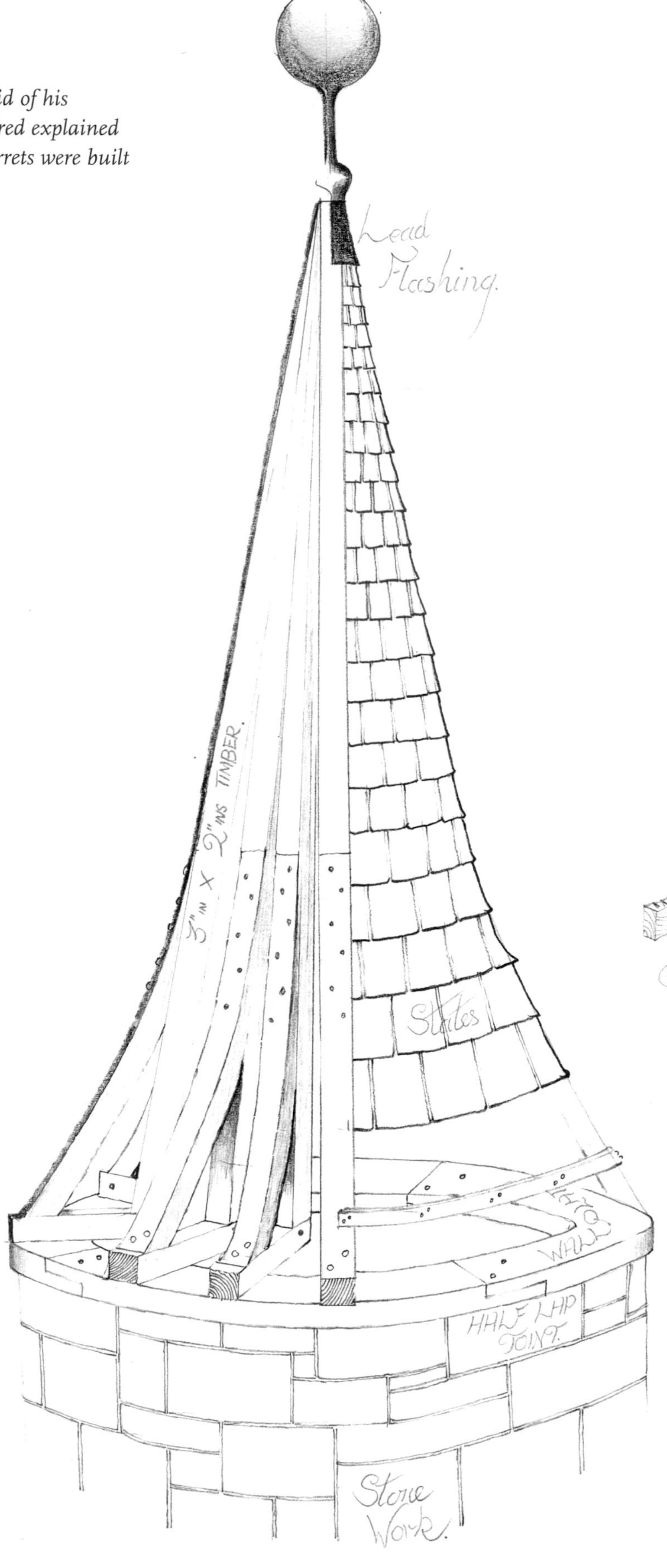

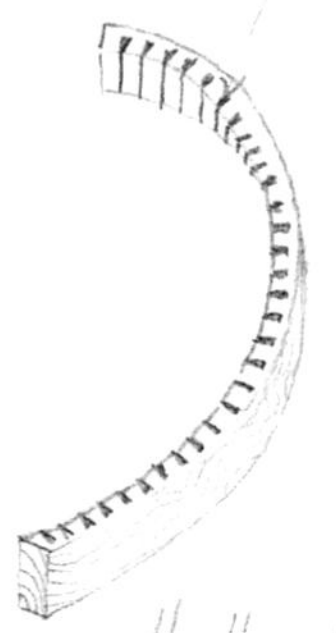

A lot of people don't realize that underneath the slates there's quite a lot of complex woodwork

Up on the roof of the castle there are lead flats surrounded by ornamental iron railings that were used by family and visitors for taking the air and viewing the surrounding countryside. But up here amongst all the pinnacles and finials and beautiful iron railings and flagpoles, the bit that interested Fred most were what he called the slated steeples. 'A lot of people don't realize,' he said, 'that underneath the slates there's quite a lot of complex woodwork that's all beautifully tapered off and rounded off. If you had to have it reslated today it'd be a fairly expensive job. Really it's a job for a good steeplejack.' Then, with the aid of his drawing, Fred explained how the turrets were built:

> This is a little drawing I've done of one of the turrets on Glamis Castle. For the basic construction of the woodwork they must have used steeplejacking technology with great baulks of timber pinned to the side of the circular part of the turret and then a few planks around that they could stand on. Once they'd got the circular wall plate rested on top of the stonework, which would be in maybe four or five pieces, they would then lift up the rafters at the end of a rope. They wouldn't be that heavy, so you could actually hold one in position while you nailed it to the wall plate. To get the slate laths to curve around the fairly tight curve as you're getting towards the top, they would make saw-cuts in the back of them. And then the slates would be nailed on in the usual manner. At the bottom where they're about seven or eight inches wide they'd have two nails and as you went up progressively as the things get smaller they'd maybe at the top only have one nail and then the whole lot would be capped with this lead finial or pinnacle. Looking at it from down below the vision is very pleasing.

To get an idea of what the castle would have looked like back in the days of the first earl you need to go into the crypt. This was the lower hall of the original fifteenth-century tower house where the lord's retainers would dine. It's one of the places that's changed the least in all the castle. The original entrance to the tower was probably always at this level, but the precise arrangements are not clear. This is one of the oldest and most impregnable parts of the castle and the massive thickness of the original tower walls can be seen by looking at the crypt windows. 'It's not been messed about with,' as Fred put it. 'It's really a wonderful bit of building; it almost reminds you of a railway tunnel, doesn't it, it's magic. It's all quite a mystery in this barrel-vaulted chamber, though, because there are two distinct lines across the ceiling where the material changes. I'm not sure why that is but it's all very interesting, I could live in here myself.'

If you had to have it reslated today, it'd be a fairly expensive job – really it's a job for a good steeplejack

Up on the second floor, the magnificent Drawing Room was once the Great Hall of the central tower, and, until its conversion in the seventeenth century, it would have looked very similar to the crypt downstairs. The first earl proceeded to convert it into the magnificent drawing room we see today. He had all the walls plastered and the fireplace done and the royal arms stuck in the middle. The second earl continued with the interior decoration, employing some of the finest plasterers of the day working in the fashionable Italian style. But all this splendour that had been brought to Glamis didn't last very long because, in 1646, the second earl died a ruined man with debts of forty thousand pounds, which was an enormous sum. It is said that he came to his inheritance the wealthiest peer in Scotland and left it the poorest. His son, Patrick, succeeded as the third earl just before his third birthday, but the estate continued to be plundered by his stepfather and he was advised that his estates were 'irrecoverable'.

It was only in 1660 when his mother died that Patrick was able to start paying off some of his father's debts and even then it wasn't until 1670 that he moved into Glamis with his wife; they had to lodge in rooms at the top of the Great Stair because that was the only part of the castle that was glazed. As soon as he took up residence he began an ambitious programme of extensions and improvements. He added a new west wing to balance the existing east wing and make it look symmetrical. Then he laid out the main avenue at 45 degrees to the original castle so that the great staircase in the corner became the centrepiece of the whole building. In front he created a baroque setting of courts, sculptures and sweeping vistas. With the structure of the house complete, the third earl turned to its decoration and for this he employed craftsmen from abroad – from Holland in particular. He built and decorated the chapel and adapted the Great Hall of the castle, which had already been made elegant with plasterwork, into an elegant Drawing Room. The painted panels on the walls and ceiling of the chapel were the work of the Dutch artist Jacob de Wet. As well as these panels de Wet did other work around the house but none has survived.

The third earl kept a diary which he called his 'Book of Record'. In it he kept details of the building works at the castle and how, as a young and impoverished nobleman succeeding to a derelict house and huge debts, he was able to overcome his difficulties. Fred was fascinated by it:

> I'm hidden away up here in the top of the clock tower where all the records for the castle are kept and I've found the third earl's diary and in it he detailed all the

A fantastic fairy-tale-castle confection of pinnacles, turrets and corbelled windows

expenses and the building operations that were going on. But it starts off with what it were like when he first came to look at it, and it says here: 'It be an old house and consequently was the more difficult to reduce the place to any uniformity.' In other words it were all higgledy piggledy and what have you. It then says: 'I did covert extremely to order my buildings so that the front piece might have a resemblance on both sides.' In other words he made it symmetrical by placing one wing on either side of the central tower with both of them coming out at right angles from it.

Not only are there all these records in this building, but also his dealings with the contractors and actual contracts that they've got and this one's an interesting one dealing with one of his main contractors. It says his lordship is very unhappy with the bill he had received from his builder. It says: 'Sanders Nisbit, as to your pretended additional work' – in other words he must have billed him for a bit extra – 'I shall receive this answer without passion. First I must tell you that I admire with what impudence you charge me any additional work.' And he goes on to say: 'If you read the contract properly, you know, you must finish the job off. But Sanders, there are a great many things to be done which are as yet not done and must be done.' But in spite of the odd disagreement, Nisbit was the main contractor for all of the work. According to another of the contracts, Nisbit had to provide five masons to work with him on site, while the Earl was to provide all materials and services of four workmen for the unskilled labour.

There have been all sorts of extensions and improvements done since the time of the third earl, especially in Victorian times. The thirteenth earl modernized and made the castle into a comfortable country home for his large family, and what we see today as we approach the castle at the end of its long avenue is basically what he created when he turned a medieval castle into a great house.

The Castle Through History

Dover Castle

Most of our great castles have been changed and added to over the centuries

Like Glamis, the majority of our great castles have been changed and added to over the centuries. Nowhere can this be seen better than at Dover. Dover Castle is one of Europe's mightiest fortresses with a history of active service that stretches from the Iron Age to the Atomic Age. The soft chalk of Castle Hill on which it stands has been shaped and reshaped for over 2,000 years into massive earthworks, ditches and mounds, to meet the changing demands of warfare. Great towers and walls have been erected on it and networks of tunnels have been dug into the chalk. Its location, overlooking the shortest sea crossing between England and the continent, has always given it immense strategic importance and ensured that it has played a major role in national history. 'It's so close,' said Fred, 'that on a clear day you can see France from the castle and they can see us. If they can see us here, this is the place they were most likely to invade, so this place is here to protect this country from invasion.'

From Iron Age to Atomic Age, Dover Castle's defences grew to meet the changing demands of warfare

The castle's shape was largely determined by the Iron Age hill fort which stood here. When the Romans came they built a lighthouse. The remains of it, perhaps the oldest standing building in England, can still be seen at the highest point in the castle. Strategically this site was so

Henry II's Norman keep stands at the heart of Dover Castle's concentric defences

important that William the Conqueror constructed the first earthwork castle here within days of his victory at the Battle of Hastings before continuing on to London.

In the 1170s and the 1180s, William's great-grandson Henry II transformed the castle. Central to the great rebuilding was the massive keep. It served as a storeroom, occasional residence of the king and stronghold during a siege. With a few modifications, it retained a military role up to 1945. Fred was interested in the way it was built:

> When you look at it you can see that the building materials they used were like an early form of concrete. It is only the mortar that holds the thing together because, although there are corners and things where they have used nicely cut blocks of stone, a heck of a lot of it is just rubble. Some people say that the mortar was tempered with ox's blood. Whether there is any truth in that I don't know, but they were damned good at making mortar. When I look at the way that it was built and the length of time that it has stood there, it gives me hope for my building work.

Henry II's rebuilding campaign, completed by his successors King John and Henry III, made Dover Castle one of the most powerful of all medieval castles. Its strength came from the double ring of strongly towered defensive walls protecting the keep, making Dover the first concentric medieval fortification in western Europe. The keep was the last line of defence and to get into it you had to get across a drawbridge from the inner bailey that surrounded it. The next line of defence beyond the keep were the walls and towers of the inner bailey with its two strongly defended gateways. As well as the gateways it is notable for its fourteen square towers. These allowed defenders to fire from the sides, adding greatly to the keep's defensive strength. Beyond the inner bailey there is an outer bailey and from the top of the keep in the centre of it all you can see ditches on both sides of the outer wall that go all the way down to the cliffs, which in themselves form an impregnable barrier for any potential attacker. But as if all that were not enough, further defences were added to the castle.

The medieval underground tunnels, which still survive today, form part of an amazing defensive system constructed in the thirteenth century. They were designed to provide a line of communication for soldiers manning the northern outworks and to allow the garrisons to gather unseen before launching a surprise attack on any invaders coming from this northern, landward end. By the 1250s the castle's medieval defences had assumed the shape and extent they retain to this day.

Standing on the top of the keep we are at the highest point of the castle, about 90 feet above the clifftop with 300 feet of cliffs down to the harbour below, which was the only seaward approach to the castle for an attacker. From here there is a good view of the Straits of Dover and the French coast beyond, whence the greatest threat always came. At the end of the eighteenth century when Britain was at war with France the castle was modernized. New defences were built which required large numbers of troops who all had to be housed. The castle became very overcrowded so, in 1797, the Royal Engineers brought in a company of miners to excavate a series of tunnels in the cliff and create an underground barracks. The first troops moved in during 1803 and during the Napoleonic Wars, the tunnels accommodated over 2,000 men.

With the development of more powerful weapons in the mid-nineteenth century and the need to protect the harbour, permanent gun batteries were placed along the cliff edge in the 1870s. The Fire Command Post for the harbour guns was established here in 1905. In 1914 this was joined by the Admiralty Port War Signal Station, which controlled shipping movements into the harbour. These two installations played notable parts in the First and Second World Wars, as did the network of tunnels under the castle. After the end of the Napoleonic Wars the tunnels had been little used. Then just before the outbreak of the Second World War in 1939, they were turned into a bomb-proof command headquarters; it was from these tunnels that the Dunkirk evacuation was directed. In the early 1960s at the time of the Cuban Missile Crisis, the tunnels were modernized and occupied again to serve as a Regional Seat of Government following a nuclear attack. No other medieval castle has undergone such a series of modernizations and adaptations to fit it for new forms of warfare.

CHAPTER TWO

PLACES OF WORSHIP

CHURCHES AND CHURCH STEEPLES HAD a special significance for Fred and he always enjoyed working on them.

When I got back home from doing my National Service in Germany, I managed to get my foot on the first rung of steeplejacking. The vicar of Bolton needed somebody to gild the weather vane on top of Bolton parish church and I got the job. It's this that launched me on my steeplejacking career and led to work on other churches. Doing this sort of work meant I had to know a bit about buildings and how they were constructed. I had to learn all about things like Gothic arches, flying buttresses, beautiful window frames, rose windows and all sorts of things that you see on cathedrals and churches that help to hold them together. I had to start finding out about things like why the top doesn't blow off a church steeple and what stops it squeezing itself apart. These are the sorts of things that I wanted to look at in the programmes we made.

Some of the cathedrals we saw on our travels took hundreds of years to build and one stone mason might have spent most of his working life working on one part of it. Today, when the time for constructing modern buildings is measured in months rather than in years, it doesn't leave a great deal of time for any ornamentation. In the olden days life was a lot slower, people didn't have computers and they didn't have big motor cars. When they got out of bed in the morning the only thing they knew really was a hammer and chisel and a big block of rock. But the skills that those workers who built our great cathedrals had, haven't disappeared completely.

Prehistory

Stonehenge; Avebury where the church overlooks the standing stones

Stonehenge

Man has been using stones to build places of worship for thousands of years but the very earliest religious structures were very different from the churches where Fred earned his living as a steeplejack. The origins of the oldest are shrouded in mystery. The great and ancient stone circle of Stonehenge is one of the wonders of the world. Set on the windswept downs of Wiltshire, what we see today are the remains of the last in a sequence of monuments erected between about 3,000 and 1,600 BC – the remains of what could be called one of Britain's first and most impressive cathedrals. It is clear that it was the focal point of a landscape that was filled with prehistoric ceremonial structures and that it fulfilled a major role in the life of the inhabitants of Salisbury Plain, but there has always been debate about exactly what purpose Stonehenge served. Its orientation in relation to the rising and setting sun has led to the view that its builders came from a sun-worshipping culture, but the mathematical accuracy of the positioning of the stones has led some experts to suggest that the circle and its banks had some astronomical significance.

Avebury has been an important place of worship for nearly four-and-a-half millennia

If the purpose of Stonehenge is shrouded in mystery what is very clear is the immensity of the task its builders faced and the ingenuity

they brought to it. Recent experiments have shown that two hundred men would have been needed to move the largest of the great sarsen stones even on level ground and these stones had to be loaded on sledges and dragged around 20 miles to the site. When they reached the site the work of erection began. To hold the uprights, pits would have been dug with one sloping side. Each stone would have been dragged on rollers so that its base projected over the pit and then by levering the other end up, the stone could be slid into the hole. 'When you think,' said Fred, 'that some of those stones weigh about 25 ton, it took some doing.' Once all the uprights were in place, they had to get the lintels on top of them. Fred's theory was that they would be raised inch by inch on a stack of timber and when they got to the right height they would be slid sideways on to the top of the uprights.

With only very basic tools at their disposal the builders shaped the stones and formed the mortice-and-tenon joints that held all the stones together. Using antlers and bones, they dug the pits to hold the stones and made the banks and ditches that enclosed them. The most important tool that survives from prehistoric times is the great deer antler pick. They may have had other, wooden tools and baskets that haven't survived, but the antler pick is a tool that is widely found on early prehistoric sites. When Fred looked at one of these with an archaeologist he found out that

Stonehenge – the remains of what could be called one of Britain's first and most impressive cathedrals

it's important not to think of it as a pickaxe. It hasn't got the weight to be used in the way we use a pickaxe. Very often when we look at prehistoric antlers that have been found on such sites, the back of what's called the coronet is heavily battered. This suggests they used some sort of mallet to drive the point of the antler into the ground and then used it as a levering tool. Fred also looked at some ox shoulder blades which have always been cited as the equivalent of a shovel.

Avebury

Stonehenge is the best known of our great megalithic monuments, but not very far away at Avebury Fred went to see one that was built on an even grander scale. Avebury rivals Stonehenge as the largest, most impressive and most complex prehistoric site in Britain. It was built and altered over many centuries between about 2,850 and 2,200 BC and the site covers a vast area. Fred was impressed by the sheer scale of it.

> They say that the huge circular bank and ditch encloses an area of nearly 30 acres. It runs for three-quarters of a mile and the ditch is about 15 feet deep. What you see before you when you come here is one of the earliest examples of building and construction work in all of Great Britain. It's one 'eck of an achievement for four and a half thousand years ago, especially when you think of all the years and the erosion there's been and the washing of the stuff back down the hole which means it must all have been a lot bigger than what we see today. We know from early excavations we're only looking at the top third of the ditch here, the rest is filled in with material that's slipped into it over the centuries.

It is thought that there were ninety-eight or ninety-nine stones in Avebury's outer circle, which is what we see today, and then there are smaller remains of two inner circles inside. Each stone stands in a pit, which is not more than 3 feet deep. They're really just balanced in position and held by smaller chock stones that wedge them into place. They are exactly the same sarsen stone as the stone at Stonehenge, but the big difference is that at Stonehenge they are shaped, at Avebury they are not. So the stones here were selected for their shape, but they were not worked in the same way that Stonehenge was. When he was at the site Fred wondered how many men were involved in the building work but found out that it is very difficult to estimate because we don't know the length of time that was spent on the work and, being so long ago, there's no written record. They've just left us with this big circular trench and the stones.

The Anglo-Saxons

Avebury has been an important place of worship for nearly four and a half millennia and as well as the pagan stone circle, there's also an early Christian church on the site. St James's dates from around AD 1000 but it was much added to in the Middle Ages. It's quite rare to find a church from the Saxon Age that has stayed completely unaltered. To see one of the few that do remain Fred travelled all the way up to Escomb near Bishop Auckland in County Durham.

Fred gets a guided tour of the Saxon church at Escomb

Escomb church

The Saxon church at Escomb is a simple, tiny church of enormous antiquity and primitive simplicity. It is one of Britain's earliest surviving churches, built largely from Roman stones taken from the nearby abandoned Roman fort of Binchester. No one knows for certain who built the church, when it was built, or why it was built in this particular location. There are no records at all. However, there are various architectural features which suggest the date of construction to be between AD 670 and 690 and it is generally agreed from the shape and style of the building that it couldn't have been constructed later than the end of the seventh century. From the outside, the outline of the building is sombre and severe. One of the things you

The tiny windows of the Saxon church are splayed to let in the maximum amount of light

Many of the earliest buildings that survive in Britain today are Saxon churches

notice is that the roof is extremely high in relation to the overall size of the church. It's not the original, which would have been thatched or possibly made from reeds.

Inside, the supports for the chancel arch have distinctive 'Escomb-style – long-and-short' stonework. The arch itself is a Roman one believed to have been brought to the site in one piece from the fort at Binchester, 'because,' said Fred, 'the Saxons didn't know how to build arches. Technically the Romans were more advanced than the Saxons even though they came before them. Really and truly the Romans were marvellous engineers.' But the Saxon builders incorporated some interesting ideas into this little church. If you look carefully, for example, you can see that the corners of the building slope slightly inwards. Windows are splayed to let in the maximum amount of light, while keeping out some of the wind and rain. The main crossbeams in the roof may date back to Saxon times, but a guess puts them not earlier than the twelfth century. What you can still see that is authentic Anglo-Saxon is some of the original plaster, which would once have covered the whole place because Anglo-Saxon churches were plastered inside and out. There are also some more features that have clearly been incorporated from the Roman fort including a Roman altar stone with a rosette on it and a barely discernible figure. This, it is thought, is the Roman god, Mithras, who was very popular with the Roman soldiers, because he was the god of courage.

St Andrew's Church, Greensted-juxta-Ongar

Many of the very earliest buildings that survive in Britain today are Saxon churches. Unlike the church Fred went to see at Escomb, however, most Saxon buildings, including their churches, were built of wood.

Walls were often clad or filled in with split logs standing vertically, but the only building in this style that can still be seen today is the Church of St Andrew at Greensted-juxta-Ongar in Essex, and here the log nave is all that has weathered the succeeding centuries. It is unclear exactly when the church was built but it is thought to be the oldest wooden church in the world. Some of the timber in the walls is from split tree trunks that were used in the construction of an earlier church on the site before the present one was built. The chancel of St Andrew's is of Norman flint, repaired with brick in the early sixteenth century.

St Laurence's Church, Bradford-on-Avon

Stone churches began to be built by the middle of the seventh century and, in addition to the church Fred went to visit at Escomb, several other fine stone churches have survived from the days before the Norman Conquest. The main features of these early churches are a fabric of ashlar or hewn blocks of masonry wrought to even faces and squared edges laid in horizontal courses with vertical joints; stumpy piers; round arches; narrow windows with splayed openings to let in the maximum amount of light; and roofs with simple barrel or groined vaults. The simple window openings we find in all the surviving Saxon churches are evidence of the difficulty Saxon builders had in forming arches. Churches are the only Saxon buildings to have survived to this day, providing us with evidence that the Saxons were masters of a graceful, simple style of building which expressed their faith in God. One of the best preserved is the little chapel of St Laurence in Bradford-on-Avon. This simple little

The arch is a Roman one because the Saxons didn't know how to build arches

The west towers and façade of Westminster Abbey were built in the eighteenth century

church was rediscovered in the nineteenth century after seven hundred years of secular use as part of a cottage it had been absorbed into. Only 25 feet long and completed around 703, the church is all that survives of the monastery of St Laurence and its sturdy plainness is the perfect expression of Saxon Christianity.

Westminster Abbey

Before the Norman Conquest, most Saxon churches were small like St Laurence. But then, in 1066, William the Conqueror defeated King Harold at the Battle of Hastings and everything changed. The first signs of what was to come were to be seen twenty years earlier with the accession of the last but one Saxon king, St Edward the Confessor. It was during his reign that the Normans became the dominant power in Europe. Edward's mother was a Norman and the future king spent his first twenty-five years in exile in Normandy. It was there that Edward first saw the Romanesque architecture that was to inspire him to create a great abbey in his homeland. The Dukes of Normandy were great builders and when Edward ordered the erection of a huge abbey church at Westminster he resolved to create a work that would match their magnificence, and modelled his Westminster Abbey on Jumièges Abbey in Normandy. The huge scale and solidity of the churches the Normans had started to build represented the idea of the Church Militant as the instrument of the state. After the Norman Conquest the face of England was to be transformed with such buildings under William the Conqueror, but the style had already been adopted in Westminster Abbey which gave a foretaste of what was to come. Edward's new church was a fine building in the Romanesque style he had come to admire in Normandy. It was vast – over 320 feet long – the largest church in the land. It was consecrated at the end of 1065, just before the Conquest.

After the Conquest the Normans began to build on a scale that had never been seen before. Huge stone castles were put in place to assert their power and authority and work began on a whole series of massive cathedrals around the country. As we saw in chapter one, these weren't just built as a tribute to God. The Normans didn't want to leave anybody in any doubt about who was in charge down here on earth so these churches were built as symbols of their authority, designed to keep control with only a handful of bishops and knights. England was changed for ever by this architecture. The old Saxon buildings were swept aside, and the way the country looked changed just as fundamentally as the way it was ruled.

R. GEORGII II VIII. A.D. MDCCXXX

The Normans

Peterborough Cathedral

When the Normans arrived in England they brought with them from their homeland all the building techniques that we see today, such as the magnificent pillars and fine arches of Peterborough Cathedral. The effect is one of solidity and permanence: arcades of round arches are carried on massive piers; walls are thick and usually made of rubble faced with ashlar. Fred said:

The main problem in any sort of building is spanning space

> They needed a massive labour force to construct buildings of this size and the Normans more or less pressganged the Anglo-Saxons into doing all the labouring. It must have been bad enough being conquered by them, never mind doing all their donkey work as well. And the work was on such a different scale than anything they'd ever done before. These were like the largest buildings in England at the time. Strength and simplicity are the main features of this style of building, which is based on very thick walls, giving the whole thing quite a chunky look.
>
> This huge expansion in the building trade led us to the building of some of our most magnificent cathedrals. One or two haven't changed since the day they were built, but the majority of them have been

The crossing of Peterborough Cathedral, one of the finest and purest expressions of Norman architecture in England

Fred's drawing illustrates the basic structure of a cathedral

They used tiers of arches to build structures much higher than they could any other way

added to and messed about with over the centuries. Peterborough is a good example of what I mean. The wonderful west front was added at the beginning of the thirteenth century, nearly 150 years after the Norman Conquest. They say it's unique in Christendom in that it has a classical-style portico like the Parthenon in Athens except that it has been built in the Gothic style with huge piers that were originally free-standing supporting arches 85 feet high. That, however, was all added later and once you get inside Peterborough you can see it's one of the finest and purest Norman cathedrals in all of England. Standing at the centre of its main crossing, with its magnificent three tiers of Norman arches, complete with chevrons and fancy bits, you get a real feeling of what Norman cathedrals are really all about. All the windows retain their original Norman shape and the view from one end of the transept to the other is one of the finest Norman views in the whole of England. The nave is largely the work of Abbot Benedict who came to Peterborough from Canterbury in 1177, and he continued to build in the Norman style at a time when his former church at Canterbury was being rebuilt in the latest French Gothic style. The result is a building which is almost entirely Norman or Romanesque as it is also known.

The cathedral is nearly 500 feet long, just over 200 feet wide and to the top of the tower it's 143 feet. It's this that distinguishes the cathedrals the Normans built from the Saxon buildings they replaced – their sheer sense of scale! The Normans managed this by importing the techniques they had learned over in France. Some of these techniques go as far back as the Romans. They used tiers of arches to build structures much higher than they could any other way. Standing in Peterborough Cathedral the whole place gives you a feeling of something permanent and solid with three tiers of rounded arches resting on magnificent stone pillars. The great columns that support all the arches are not built of solid masonry, but as tubes filled with rubble. A tube has got more rigidity. It's also lighter and it's quicker, cheaper and easier to build.

When you look around on the surface of some of the stones, there are some interesting masons' marks on them. Modern stonemasons even use them to this day, and in a way it's a signature of the man who actually made that particular stone. And if there's any rough workmanship, I suppose they could nail him. The thing is that when the place was built you wouldn't have been able to see any of these marks because the whole place was heavily limewashed and painted.

Fred's drawing of a Norman arch with chevrons

Norman arches, also known as Romanesque, at Fountains Abbey

The Arch

The main problem in any sort of building is spanning space. In medieval times this was almost exclusively achieved with the arch. The most commonly used method was to use wooden 'centring', which was the name given to the structure constructed to support the arch as it was being built. Fred explained:

The Normans built with semi-circular or round arches, just like the Romans used to do. And of course that's why sometimes they're called Norman and sometimes they're called Romanesque. The arch really is the main thing about all these cathedrals and it did three things. It saved material. It also looked very attractive and it let lots of light flood in from the sides. Lots of people wondered how they built arches. Well, it's very simple really. They actually built a wooden framework or centring in the shape of a curve that matched the inside of the arch. This is one of the reasons why woodwork was so important in buildings like this. The carpenters didn't just do the decorative carving but played a large part in the actual construction of the building. Anyway, once the carpenters had done their work and they'd got the centring in place, the masons would come along and, starting from the bottom at each side, individual stones or bricks were built round the frame and they worked their way to the centre. The keystone, which is the central or topmost brick or stone in the arch, was then placed in the middle. Then you take the frame away and if it's been built right the arch will stay in place. The more courses of brick or stone you build on top and the more weight that goes on it, the more solid the whole thing becomes.

The medieval workmen may have used a number of methods for lowering the centring – or 'decentring', as it was known. One was to have the centring supported on poles set in sand-filled drums. When decentring was to take place, the base of each drum was unplugged, the sand would run out, and therefore the wooden centring would lower by a few inches. But the most common way was to use wooden wedges and knock these away. In medieval times the main cost in building was the materials and not the labour. Therefore the scaffolding, or 'falsework' as it was known, was kept to a minimum. One way of doing this was for the centring to be supported on a wooden framework which was often mounted on wheels. This would mean that after the centring had been lowered the wooden framework could be moved along to construct the next arch in the vault. Also, by using the same wooden centring the arches would be almost exactly the same shape. Fred gave a practical demonstration of arch-building and explained:

> An arch is a really efficient way of carrying weight, because it spreads the load of the building equally across all parts of the archway. So, if you build an arch and then put it on top of another arch, all the weight on the upper arch is spread across the lower arch. Then if you put another arch on top of that one, all the weight is spread again, and you can end up building something much bigger and heavier than you could if you just stuck to the old Saxon way of building. Of course, you can't just keep on doing this. There's a limit to the amount of weight an arch can take. And, the thing is, if you are going to build something that is made up of stone arches, you need really thick walls at the foundations, to take the weight.

Ely Cathedral dominates the flat Fenlands that surround it

At Peterborough Cathedral there are three tiers of arches all along the nave and across the transepts – all pure Norman.

During the time the great Norman cathedrals were built, there were great changes in the building industry. New methods and ways of doing things were being introduced. The medieval periods and styles overlap and although the Norman style persisted until around 1190, the achievement of the latter part of the twelfth century and the thirteenth century was to transform the heavy Romanesque style into a marvel of lightness and flexibility. Rib vaulting was a new invention that emerged at this time and it was to revolutionize the way the medieval cathedral-builders could go about their work. A good example of early rib vaulting can be seen in the side aisles at Peterborough, which was one of the first places where they actually used it after its invention at Durham Cathedral only a few years earlier.

The Rib Vault

Until this time most naves had flat ceilings, like the remarkable painted wooden ceiling that still survives at Peterborough. The roofing of a great space in stone was one of the ambitions of twelfth-century architects. The earliest method used was probably to fill in the room to be vaulted with earth. At the top of the room the earth would be moulded into the formwork to give the vaulting its shape. There were a number of problems with this method. Firstly the earth put pressure on the walls and often caused them to collapse. Secondly, it meant that no other work could be done in the room because it was full of earth. Barrel vaults and groin vaults were not suitable for large interiors because they involved great masses of masonry resting heavily on solid walls. By conceiving the rib vault at the end of the eleventh century the masons of Durham solved this problem and paved the way for the great constructional and aesthetic achievements of Gothic architecture. The ribbed vaulting of the massive roof at Durham was considered a construction miracle of its day.

The ribbed vault concentrates force on to a single point. It's basically a framework of diagonal arched ribs which carry the cells or surfaces of a vault. These cells cover in the spaces between the arches, which take the weight of the walls above and give added support to the building. This way of building was important for the development of the Gothic cathedral with its pointed arches where the whole building became a kind of rib vault. Durham sees the first appearance in Europe of the pointed arch; the feature that was to be the secret of the world's greatest works of architecture over the next four centuries.

Ely Cathedral

Ely is an enormous fortress cathedral, over 500 feet long and 200 feet high, which took thirty-seven years to complete. The magnificent west tower is almost like a Norman keep, complete with battlements on the top, and it's over 200 feet high. This huge structure dominates the whole area and from the top of it you can see for miles in all directions across the flat Fenlands that surround it. 'You hardly needed a castle when you had a cathedral like this,' Fred said. Like Peterborough, Ely is one of our best preserved Norman cathedrals. The foundations of the magnificent nave were put in around the year 1100 and it was finished in 1189. Fred pointed out some of the features of the building work:

> One of the interesting things when you look at it is the much cruder finish up near the top which indicates that this was where the junior masons were allowed to try out their skills out of the way where they thought nobody would see their work properly. This main part of the cathedral was built over a period of a hundred years and you can see how the way it was built changed over this time as the Normans moved on from the round arch and started to build with the more pointed, Gothic arch. Really the big difference between the Norman arch and the Gothic or the pointy arch is the fact that with the rounded Norman or Romanesque one, the thrust went sideways. This meant you needed much greater weight in the abutments or the walls of the building that the thing were in. With the pointed arch all the weight went straight down and there was very little pressure sideways. If you go into some of the Gothic-style cathedrals like York Minster and look how slender the pillars that support the pointed arches are and then you go in a Norman one like Ely or Peterborough and see how chunky everything is, you can see why it were regarded as a great advance in architecture.

The Gothic Cathedral

At Chester Cathedral, a Norman and Gothic arch stand side by side

A good place to see the difference between Norman and Gothic at close hand is Chester Cathedral. The first building on the site that is today occupied by the imposing red sandstone cathedral was a Saxon minster. Following the Conquest Norman monks came to Chester and began work on an abbey, but the building process was slow, taking about 150 years. The first work they did was in the heavy Norman style, but by the time they got to the later stages it had evolved into the lighter, more elegant Gothic style. If you go into the North Transept you can see very clearly the difference between a Norman and a Gothic arch, because here the two stand side by side. A weathered Norman arch dating from about 1092 with an arcaded gallery above it adjoins a Gothic arch which appeared in the fourteenth century when the nave was rebuilt. 'It looks like a real mixture,' Fred said when we went to see it, 'but they didn't seem to mind how it looked. You can see how much thicker the walls are round the Norman arch. I suppose the reason for that was the semi-circular arch needed them to be thicker to take all the outward thrust, as against the Gothic job where all the weight was downwards. This was a great step forward because you could do more with a Gothic arch. You could go much higher and

The lantern at Ely Cathedral: a masterpiece of medieval engineering

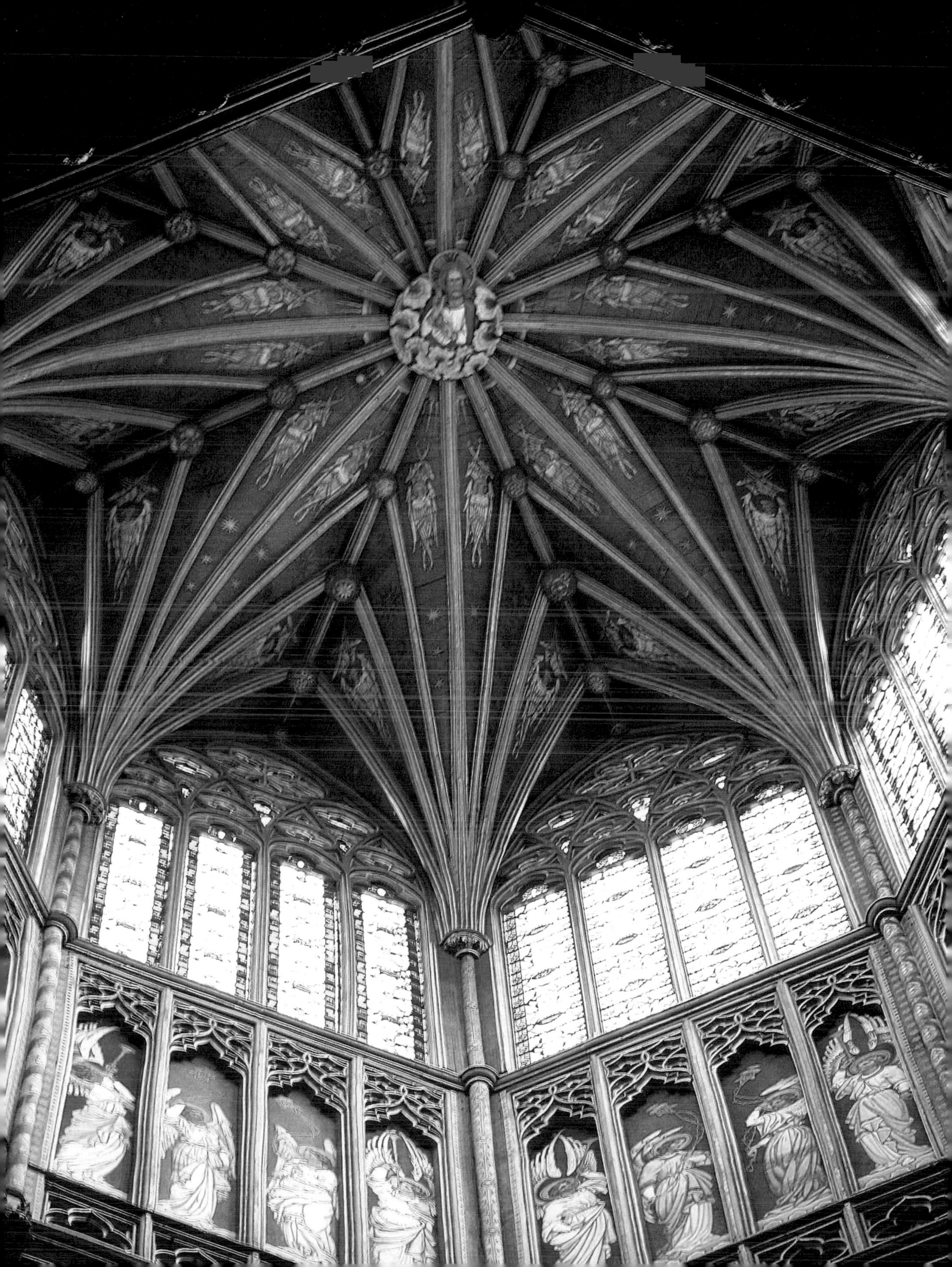

This was a great step forward because you could do more with a Gothic arch

you could vault over bigger spaces. It was much more flexible.'

The twelfth to the fifteenth centuries was the great age of cathedral-building in Britain and we have been left with some magnificent monuments to the skills of the medieval builders and stonemasons. Great cathedral towers came to dominate the skylines of cities that were anxious to display their wealth and power and it was only stone construction that allowed the possibility of erecting taller and taller buildings. Great advances were made in building techniques and by the end of the twelfth century a distinctive new style began to emerge – buildings full of space, air and light, with the typical ribbed vaults and pointed arches of the new Gothic style.

The Flying Buttress

The introduction of the pointed arch enabled church- and cathedral-builders to develop the elaborate system of vaulting and buttressing which is the basis of Gothic architecture and the thing that differentiates it from the earlier Romanesque. One of the great engineering advances of the time was the introduction of the flying buttress. Essentially a buttress was a mass of masonry or brickwork projecting from or built against a wall to give additional strength, usually to counteract the lateral thrust of an arch, roof or vault. The flying buttress became a distinguishing feature of many of the great Gothic cathedrals of the Middle Ages. Basically it was a half-arch which transmitted the thrust of the roof across an intervening space to a buttress outside the building. The space it crossed might be an aisle, a chapel or a cloister. As a result the buttress flies through the air, hence its name. The highest part of the arch supports the upper part of the wall and the base of the arch is mounted on foundations or pillars or, sometimes, on other buttresses. The employment of the flying buttress meant that the downward and outward thrust of the vault, or arched roof of the cathedral, was met by the equal and opposite upward and inward thrust of the flying buttress and the upward thrust of the columns. The result of this was that the main load-bearing walls could contain cut-outs for large windows that would otherwise seriously weaken the walls of the vault. When you look at a flying buttress you will always see a pinnacle on the top. This is not merely decorative. It is there because the effectiveness of the flying buttress is a function of its shape and its weight and the pinnacle adds extra weight. Fred said:

> Most people go to cathedrals and look at flying buttresses and they think maybe they're a bit ornamental, but what a lot of people don't realize is that they're very

The flying buttress is a half-arch which transmits the thrust of the roof to a buttress outside the building

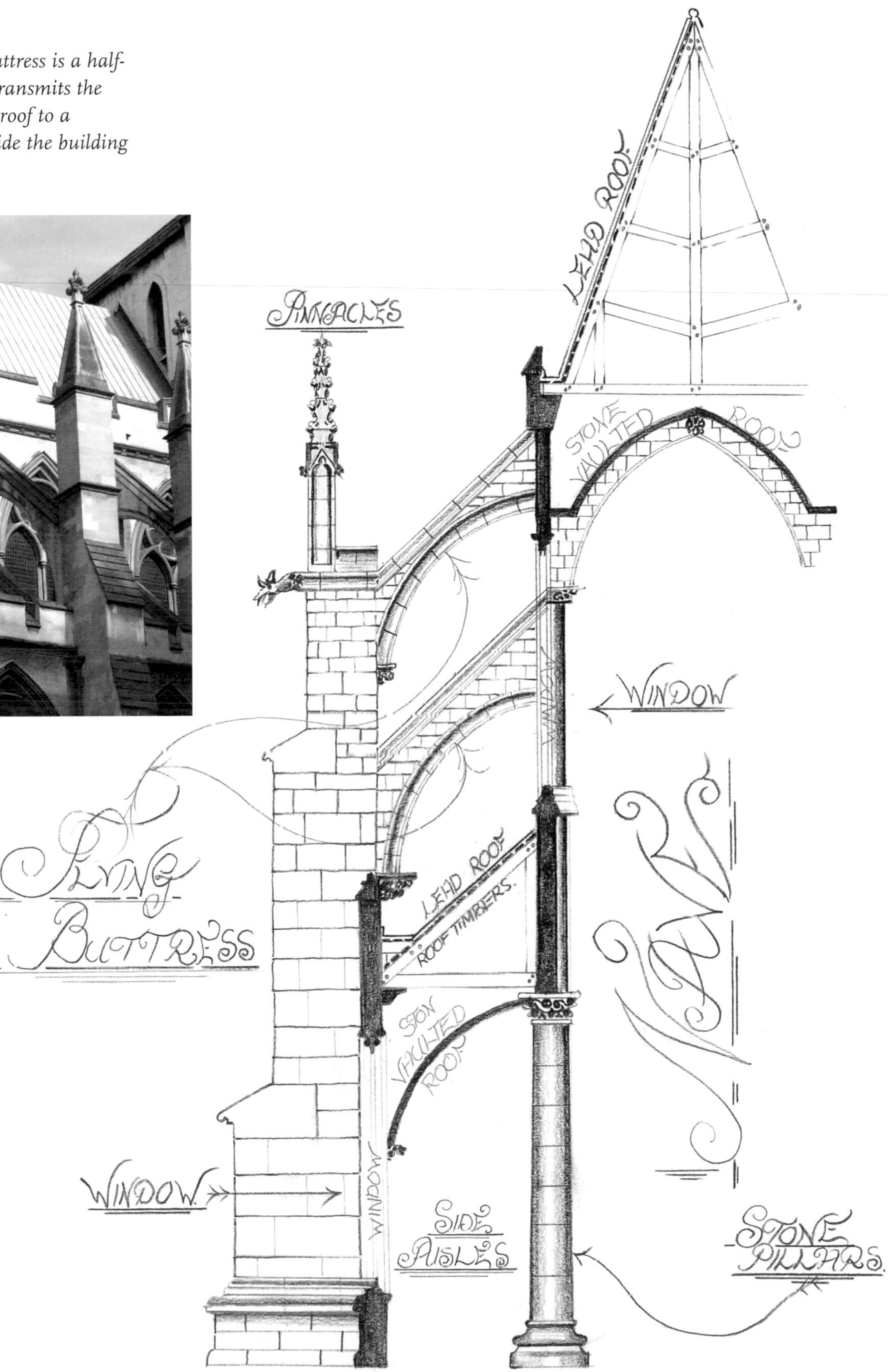

Carving a finial for the restoration of the Great West Door, York Minster

important. When you go in these great cathedrals and you stand in the main aisle and look down and you see all these great big windows, you think how wonderful, eh? But why doesn't it all collapse when you see two great rows of pillars with arches on top and then great tall leaded windows up to the ceiling nearly, and then beautiful groined stone roofs. In actual fact when they invented the side aisles with groined roofs they built the big buttresses outside, out in the rain, and they put as much weight on them as they could so they could build these little half-arches – the flying buttresses that hold together the solid bits in between the windows.

Flying buttresses, along with pointed arches and ribbed vaulted roofs, enabled the builders of Gothic cathedrals to construct roofs that were not only higher but wider than had been possible before.

The techniques for building flying buttresses were very similar to those employed to build arches and ribbed vaulting. Centring would be used to support the stones and help maintain the shape of the half-arch until the mortar was dry.

The Medieval Building Site

To build structures like this, you needed really skilled craftsmen – not just stonemasons, but carpenters too. And they

When most people think of cathedrals they think of stonemasons, but there's a bit more to it than that

didn't just build the scaffolding to hold up the arches as they were being built. They built all the accommodation and storehouses that surrounded a project like this, which could take decades to complete. And they did all the woodwork that decorated the interior of these cathedrals. It was the great age of cathedral-building that created master-carpenters capable of pushing the boundaries of their craft to new limits.

Fred started his working life as a joiner and he reminded us of the importance of his first trade in the building of our great cathedrals.

> When most people think of cathedrals they think of stonemasons, but there's a bit more to it than that as there were as many joiners. This was because, in spite of the fact that the cathedrals were built of stone, wood was indispensable for the construction of the framework of the roof, arches and groining and for all the beautiful woodwork you see inside a cathedral.
> I would rather think that just like the stonemasons the men who worked in wood would come into various categories. The carpenters did the rough stuff like all the centring for the arches and the joiners did the finer bits like all the beautiful wood carving you see in a cathedral.

The carpenters, however, played another very important role on a medieval cathedral building site because they were the men who built all the scaffolding and platforms that the stonemasons worked on. They also constructed the lifting machines and equipment like the windlass we saw at Peterborough. Without this medieval machinery the stones couldn't reach the tops of the buildings as they grew higher.

The vast majority of the workers on a cathedral building site, however, would be the stoneworkers and these, said Fred, fell into two categories.

> There were the stonemasons who did all the lovely tracery for the windows and things like that. These were the stone-cutters. Then the other branch of that trade would be the rough guys who did the in-fill in the walls and made the mortar. Basically, the stonemason's craft involves two types of work. There's a sort of geometric masonry which is very disciplined and forms definite lines. Big pieces of stone are cut with straight edges and these are used as the main building blocks. The men who did this were the setters and wallers and they were skilled in heavy ashlar work. But if there are any embellishments to do like faces or flowers or leaves, these blocks go into the carvers' workshop. The men who did this work were

Fred's practical demonstration of the way a vaulted roof was built

Ribbed vaulting at Fountains Abbey

the freemasons and when you look closely you can see that the whole of a building like Ely is covered with their intricate handiwork.

The thing that made all this fancy stonework possible was the rapid improvements that were being made at this time in metalwork, especially in the blacksmithing department. They made better tools with better cutting edges which enabled stonemasons and the joinering department to do much finer work. All the fancy tracery and everything were much easier worked with better steel in the tools. It enabled the joiners and carpenters to make really graceful centres for building all those beautiful groined ceilings. Now that more resistant types of stone and more durable wood could be used, the cathedral builders could design columns that were narrower and more graceful-looking. And sculptors and carvers were able to do finer and more delicate designs.

At Ely Cathedral Fred gave us a picture of what the medieval construction site would be like:

On the grass outside the cathedral at that time it would all be a great hive of industry and there would be quite a few wooden sheds that the craftsmen had made themselves to protect themselves from the rain and the weather and the elements while they went about their work. It would have been a bit like the castles; they'd have building seasons when the sun came out in the summer and they'd all be happy and crashin' away up there on the walls but in winter I suppose they spent most of their time underneath a roof

> down on the grass, chiselling beautiful tops for columns and things of that nature. It must have all been very pleasurable.

The wooden sheds or 'lodges' as they became known were primarily a place of work and an area where tools could be arranged and stored. The working day was regulated by the ringing of the lodge bell and work was carried out from daybreak until sunset. During the winter, when the laying of stones had to be stopped due to the risks of ice and frost, the lodge would lay off almost half the workers. Whenever the building site was closed for any length of time because of bad weather in the winter they became training centres where the master and craftsmen taught their art to the young apprentices. Masons worked long hours. Generally whenever there was daylight they would be at work and this meant that they earned substantially more in the summer than they did in the winter.

One of the things Fred found particularly interesting when we were at Ely was the effect that alterations and additions made by later builders had on the original Norman structure, because not all of Ely dates from the time of the Conquest. It was extended and added to for centuries afterwards, and the unique thing about the cathedral is that you can see the development of Norman architectural styles at the time of its building from the crude rounded arches of the eastern transepts, through the classic Norman nave to the beautiful Romanesque work near the west tower and the Early English Gothic pointed windows at the top of the South-west Transept, added before the end of the twelfth century. Then there are

The tower at Ely Cathedral: the top 60 feet were added 300 years after the Normans first built it

the changes that were made to the tower and, being the highest bit, it was one of the parts that interested Fred most:

> The tower is even taller now than it was when the Normans first built it, because three hundred years later they erected a magnificent, 60-foot-high octagonal-shaped bell chamber on top. It sits there like a crown on the old Norman structure dominating the landscape. The only problem was that those later medieval engineers didn't do their calculations very well because the alterations to the tower have put a massive extra weight down on to the old Norman foundations. They realized that the original walls weren't strong enough to take all this extra weight, so they very cleverly put a sort of stone skin inside the original tower and reinforced the arches to take the strain. But it's been a bit of a headache for successive generations of architects because its weight is clearly too heavy for the Norman foundations.

In spite of that, between the twelfth and the fifteenth centuries, master craftsmen constantly pushed the boundaries of their skills to new limits. One of their greatest masterpieces is the lantern at Ely Cathedral. Fred explained:

> It came about because of a bit of a disaster they had there. In 1322 the central tower of the cathedral collapsed and this destroyed the Norman choir. But instead of rebuilding the conventional four-square tower, the cathedral bursar Alan de Walsingham produced a design that made the crossing octagonal by adding a diagonal bay across each corner. It was a breathtaking feat of engineering that began with the building of eight huge stone pillars over 100 feet high. The octagon took six years to build but the biggest challenge they faced was that the roof over this space needed to let in the light; and the solution they came up with was the lantern. To construct the lantern in stone would have posed so many problems that they called in from London

Fred demonstrates how the lantern at Ely was built

William Hurley, King Edward III's master carpenter, and asked him to devise a solution in timber. The lantern he designed doesn't rest on the vault beneath it, which is also of wood, but on a cantilevered wooden frame which is hidden from view. To build it huge timber triangles were constructed at the top of the eight supporting pillars. At the apex of each of these triangles are the vertical posts of the octagon gallery, each 20 feet high. These support a ring beam on which stand the 40-feet-high posts of the lantern. It all weighs 200 tons and has a perfect perpendicular downward thrust, making it one of the great masterpieces of medieval engineering. Light floods into the octagon not only from the great windows in the side walls but also downwards from the lantern. The whole construction, which took fourteen years to build, is one of the most daring and imaginative pieces of cathedral architecture in England. Really, when you think it weighs 200 tons and it were done all them years ago, it's a credit to them men. A lot of them couldn't even read or write, but they did it somehow or other and all done for the greater glory of God.

By the time the lantern at Ely was built the transition had been made from the Norman to the Gothic style of architecture. This new system of flying buttresses and vaults with crossing ribs enabled the great medieval cathedrals to have large windows, through which light would flood into vast naves. Stained glass had come into use during the Romanesque period but it was not until Gothic times that its full development occurred.

The nave at York Minster is the widest Gothic nave in England

York Minster

York Minster is the largest English medieval cathedral and the second largest Gothic cathedral in northern Europe. The present minster is the latest of several to stand on the site. It was in 1220 that the present building was begun and it took 250 years to complete. The nave, which was begun in 1291 on the Norman foundations, is unusually tall and broad, making it the largest medieval hall and the widest Gothic nave in the whole of England. It has 100-foot-high limestone pillars and, because they were worried by the potential weight of stone vaulting, the builders spanned the space with a ribbed vaulted roof made of wood painted to appear like stone. The aisles have vaulted stone roofs.

The octagonal-shaped Chapter House was begun in about 1260. It is unusual, given its size, in that it doesn't have a central column to support the roof vaulting. Around its walls are some of the minster's finest carvings, 80 per cent of which are original, dating from between 1270 and 1280. The Great East Window, built between 1405 and 1408, is one of the largest areas of stained glass in the world with dimensions similar to those of a tennis court. Because of its vast area, the window is supported by extra stonework which forms an internal screen and within this there are two walkways which go right across the face of the window.

In 1984 a fire destroyed the roof of the South Transept, so the skills of the minster's work force were needed to restore it to its former glory. As a result of the fire, the roof of the South Transept had to be completely rebuilt and it was decided to use the traditional construction method. Oak ribs were used to form the shape of the roof and these were locked into shape with bosses. As only six of the original bosses were sufficiently unscarred to be reused, sixty-eight new ones had to be designed and carved. Fred marvelled at the exterior:

> When you look at the stonework on the outside of York Minster, it's amazing really the amount of detail you can see – all the mouldings, the niches, the beautiful statues, the window openings, the tracery in the windows. The detail around the Great West Doorway is beyond belief, absolutely marvellous. To do all that work and to keep it all looking as good as it does, has to be, without any shadow of a doubt, one continuous effort that will go on for ever just to keep the thing standing up. I bet they have been working on it every day since it was built. People say it's a pity they can't do things like that now, but I'm afraid they're wrong, because most of what you see today up above the doorway was made in a yard, just round the back of the minster. I went to

have a look in there to see how they do it and found out that it's all very much the way that it would have been done back in the Middle Ages. Before they can start doing any chiselling away at the stone, detailed drawings have to be done just as they would have done then. From the drawings they make a template which the stonemason uses to carve out the basic geometric shapes. This is the first stage – the geometric masonry which is very disciplined following definite lines in the stone. Some of this goes straight into the building, but if there are any embellishments to be done in the stone with flowers or leaves or anything like that, then it goes into the carvers' workshop. There is a young lady there who was making the most beautiful finial. I had a good chat with her and asked her how she went about starting it off and, as a result, if I were asked to make one now, I would have a much better idea of how to go about it. She had been chiselling away at this one particular piece she had been working on for eleven weeks. Just think about it, every day for eleven weeks for one piece and it was only about 2 feet tall. But think of all that wonderful feeling of satisfaction when you get to the end of it, when you consider that you started with a pyramid-shaped lump of rock, 2 feet tall, 18 inches at the bottom and 1 inch at the top, and then you transform it into a beautiful thing. Yeah. When people walk by York Minster they don't appreciate all that great effort that's gone into it, do they?

English Baroque

St Paul's Cathedral

Just over a hundred years after the Reformation and the Dissolution of the Monasteries there rose in London a magnificent new church that didn't have any of the austerity of a medieval monastery. 'St Paul's Cathedral, designed by Sir Christopher Wren,' declared Fred, 'is something else. In its day it must have dominated the London skyline, but it's a shame that nowadays you can't really see it properly for all the modern buildings that have sprung up around it.' A cathedral had stood on this site since the early years of the seventh century and the original Saxon church had been destroyed by fire and rebuilt on a number of occasions. After the last disaster in 1087, the Normans constructed a massive church whose size and style reflected the importance of London, the capital of their newly conquered kingdom. The cathedral they built was the largest church in England and the third largest in Europe. It had the tallest spire and steeple ever built in England, but the spire was made of wood and it was burnt down on a number of occasions when lightning struck it. Then in 1666 came the greatest disaster of all when the Great Fire of London put it completely beyond restoration.

When we filmed at St Paul's, they were in the middle of a big programme of restoration work

Sir Christopher Wren's crowning achievement – the dome of St Paul's

The man chosen in 1668 to design and construct a new St Paul's was Christopher Wren. By education this greatest of English architects was not an architect at all, but a classicist, scientist, astronomer, engineer and mathematician. He was also a deeply religious man. Early in his career he began to bring the precision of mathematics to the solving of architectural problems of structure and space and the brilliance of his mind allied to an uncommon ability to solve practical problems soon brought him to the attention of King Charles II. Wren travelled to France to study buildings but it was the Great Fire that brought him to the architectural fore back in England. He was appointed by the king as one of three commissioners to survey the extent of the damage and to advise how the city should be rebuilt. A tax was put on coal to pay for the rebuilding of more than fifty churches that had been destroyed in the fire and it was this that financed the building of St Paul's.

Inspired by the domed churches he had seen in France, Wren had the idea for a dome resting on eight arches for the new St Paul's. His first design for the cathedral was rejected, as was his second, on the grounds that it was untraditional, but the king liked his third design and issued a warrant of approval to allow the project to go ahead in 1675. Then came one of the bits that got Fred particularly excited:

> Of course before they could build it they had to get rid of the remains of the other one and believe it or not he tried explosives, small charges of gunpowder to help topple the massive tower pillars which stood over 200 feet high. But he upset the neighbours with all the noise, so they had to resort to a battering ram to get rid of the remains of the pillars. When the remains of the old cathedral had been demolished and the site was cleared, work on the new cathedral could

The highly ornamental façade of St Paul's – a characteristic of the English Baroque style

begin. Wren himself insisted on supervising the preliminary measurements. Throughout the thirty-five years of building work, he supervised and engaged the finest craftsmen, scrutinized and signed the accounts and visited the site each Saturday. To make sure there was no interference from anybody, he insisted that the work be carried out under conditions of some secrecy, with whole sections of his great design being kept under wraps as the building went up.

What Wren achieved during those thirty-five years is now one of our greatest national monuments; St Paul's Cathedral represents inspiration, beauty and craftsmanship on a grand scale. The architectural style of the church is English Baroque. The Baroque style was one of great grandeur that was widely used in Europe during the seventeenth and early eighteenth centuries. It is associated with the Roman Catholic Counter-Reformation and it is characterized by highly ornamental façades, exuberant decoration, expansive domes and curves and a sense of openness and space, but it was a daring and controversial choice for an English cathedral. In its purest form it is to be seen in the great churches and cathedrals of seventeenth-century Italy. When the style was adopted in England it was tempered by classical elements and the great west front of St Paul's, where the cathedral is entered, has two storeys of dignified classical columns flanked by twin towers.

The interior of the cathedral is elegant, flooded with clear light through the plain glass windows which Wren preferred. When you enter you look down the nave towards the cathedral's crowning achievement: the dome. This is among the largest and most majestic in the world, comparable with St Peter's in Rome. The main structure is of Portland stone from Dorset. Eight piers support its 65,000-ton weight and eight arches spread the load. Above them is the Whispering Gallery, which gets its name because a whisper against the blank circular wall can be heard on the opposite side some 136 feet away. Above this are the Stone and the Golden Galleries. When the dome was being built, Wren was hauled up in a basket two or three times a week to see how the work was progressing. The central pavement area underneath the dome is decorated in a compass design in coloured marble. On it are the words: *Beneath lies buried the founder of this church and city, Christopher Wren ... Reader, if you seek his monument, look around you.* 'Never,' said Fred, 'was a truer word engraved in stone.'

The dome is a brilliant piece of engineering and it is only by climbing up into it and exploring it that we can fully understand

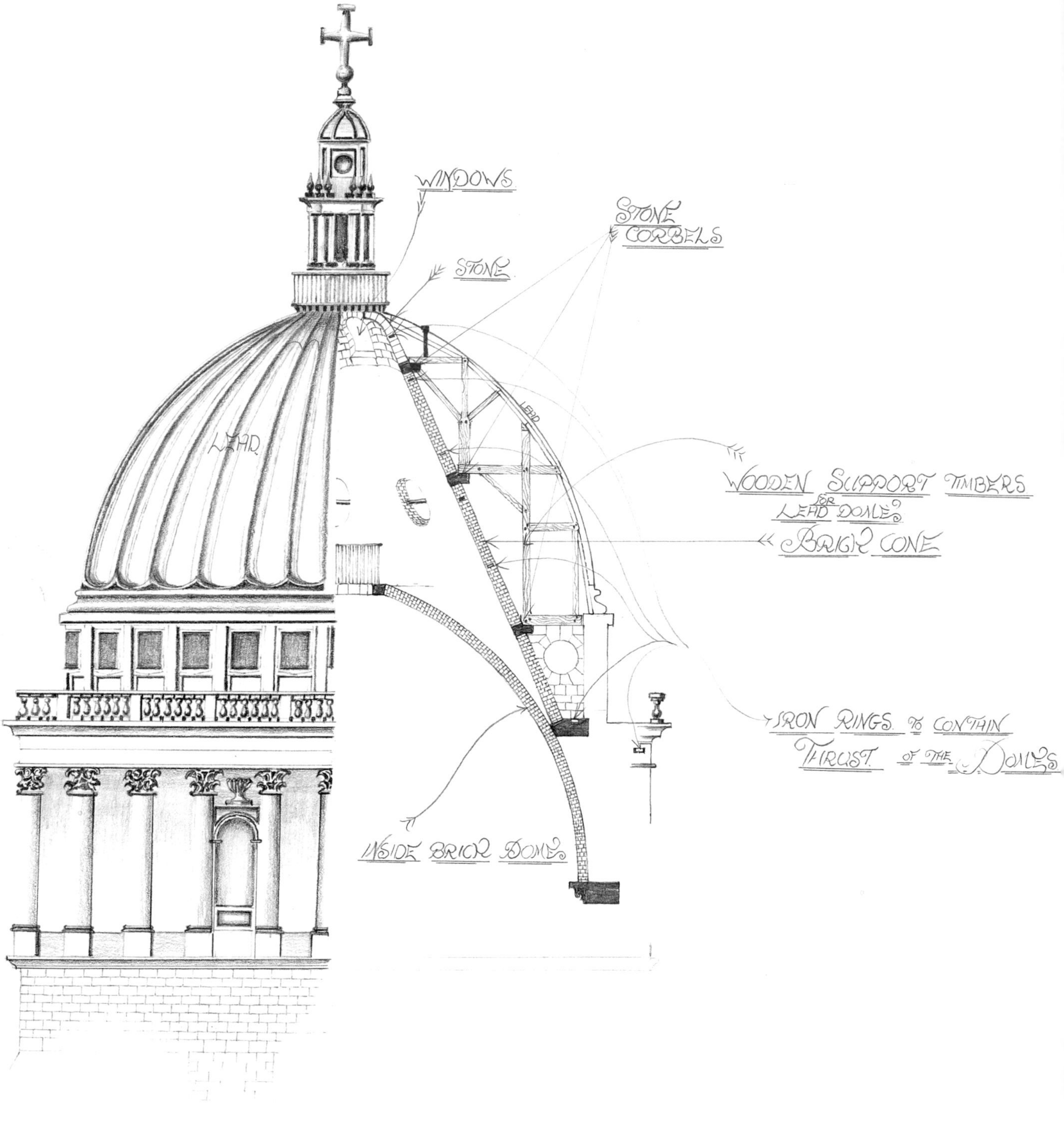
WINDOWS
STONE CORBELS
STONE
LEAD
LEAD
WOODEN SUPPORT TIMBERS FOR LEAD DOMES
BRICK CONE
IRON RINGS TO CONTAIN THRUST OF THE DOMES
INSIDE BRICK DOME

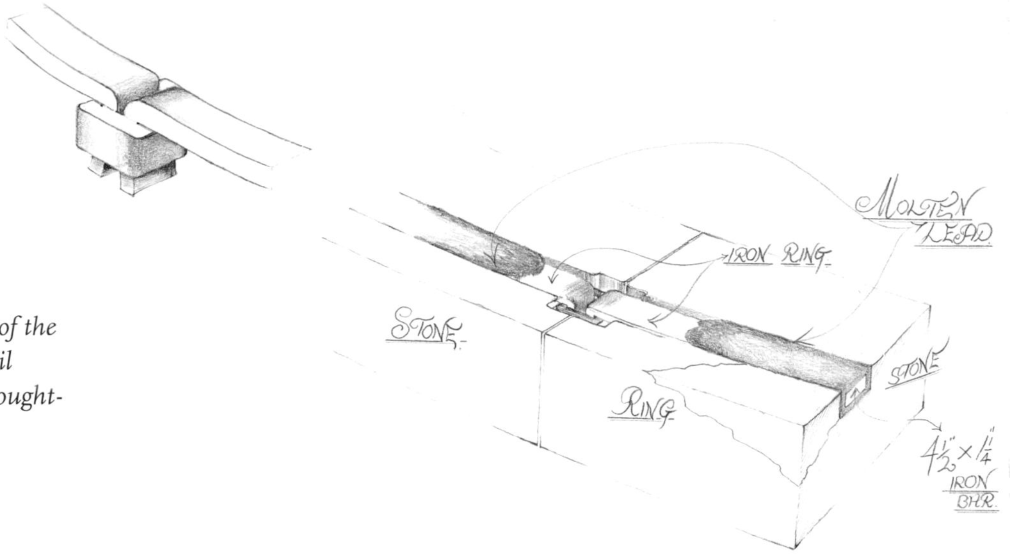

Fred's drawing of the dome with detail showing the wrought-iron chain

Wren's inventiveness which combined his skill as an architectural designer with engineering. On close inspection we can see that it actually consists of three structures, one within the other. Fred described it:

> St Paul's Cathedral is really a wonderful building. When you stand there in London and look at it you imagine it's just a big dome but it isn't really, there are actually three domes one inside the other. When you are standing down in the bottom, you can see a beautiful dome all hemispherically shaped when you look upwards, but on the top side of that there's a big cone made of brick and stone that nobody can see unless you climb up into the dome. When you look up from the crossing at the centre of the cathedral you can see that there are great stone corbels that support the framework of this cone. Then stuck to the outside of the cone there's a load of timber roof trusses that hold yet one more dome that is sheathed in lead. This is what you see from outside.
>
> When you look at my drawing you can see that the greatest area of thrust of the whole lot is on top of the eight great pillars. But with all this pressure, how do you stop the whole thing pushing out on itself and collapsing? Well, Wren came up with an ingenious solution. He put a great wrought-iron chain all around the dome and sunk it into the band of stone that went all around the perimeter just at its base where it meets the main roof of the church. What this does is to take the strain of the outward pressure of the dome. When I had the pleasure of visiting St Paul's they were in the middle of a big programme of restoration work on the dome and we were able to see this iron band. I went all the way to the top of the dome to see it and when I was standing out there I could count no less than two hundred tower cranes in that big city. But nothing that was being built was as impressive as St Paul's Cathedral and its dome. When you see that stone on the outside of the dome close up as I did, there is a lot of evidence of repairs that were done

The bit I really like is the steeple, the lovely, graceful-looking spire – it's a work of art, you know

in days gone by, but the thing I really wanted to see was the iron chain that Wren had sunk into the stone. I had a good look at the work they were doing up there but even now you don't see much of the chain. All you can see of it are the bits they have got exposed in the section they are working on, but it does reveal Wren's engineering that keeps the dome standing. The trouble is that over the years damp has started to seep through the stonework and corrode the metal which has then expanded and started to crack the masonry and what they were doing was putting that right. I was very pleased to see that the gentlemen who are doing the restoration work are making a very good job of it and I think Sir Christopher Wren himself would be highly delighted if he could see their work.

When you think how old St Paul's is, it really is a great credit to the man who designed it so well. For its age it is a magnificent piece of engineering. One thing I noticed when I was there is that the architecture of St Paul's is almost identical to Bolton Town Hall. There are some great pillars on the outside and the balustrade around the top and the cornice moulding are almost identical. I think the guy who designed the town hall must have been down and had a good look at St Paul's to get some ideas. The Victorians used to copy off everybody; they copied off the Greeks and Romans and from everything else that had come before them. They would copy all sorts of nice bits and pieces if they were good to look at.

The Marble Church

It's interesting that over the centuries many of our best architects, builders and craftsmen have reserved their finest work for places of worship. However, a church doesn't have to be big to be beautiful. Fred recalled:

On a recent job in Wales we passed a magnificent marble church. It's actually called St Margaret's Church and it's at Bodelwyddan, which is near Rhyl in North Wales, but it is more commonly known as the Marble Church on account of all the different varieties of marble used in its interior construction – fourteen different varieties in all. This finely ornate church is in the decorated Gothic style and there is an amazing richness of design to be found in it. In the nave, the pillars have shafts of Belgian red marble and the capitals are of richly carved stone while the chancel and the sanctuary steps and flooring are of Sicilian marble. On the hammerbeam roof there is not a single screw or nail used in its construction; every junction of the timbers is secured by pegs. There are also some very fine examples of the craft of the woodcarver to be seen there.

But what impressed Fred most about the church was the spire. 'The bit I really like,' he said, 'is the steeple, the lovely, graceful-looking spire. It's a work of art, you know.' The tower and graceful spire of the Marble Church can be seen for miles. The spire rises to a height of 202 feet, and its four traceried windows, bands of ornamental tracery, finely worked buttresses and the carved portraits at every corner of the four pinnacles or finials make it unique. 'The whole spire,' said Fred, 'stands on just eight stones, one on each corner, and it always amazes me how the man who designed it knew enough about the strength of the materials that he used for building it that he was able to calculate that the weight on those corners wouldn't crush them, resulting in the whole thing tumbling down. He was a man who knew the rock that it was built out of. Either that, or he was one hell of a gambler.'

When we filmed at the Marble Church Fred had the chance to climb up into the spire that he had admired from ground level for so long. 'Ever since I passed by here years and years ago and stopped one day and had a look at it,' he said, 'I always wanted to have a closer look and get inside it like where we are now.

The spire of St Margaret's Church, Bodelwyddan, North Wales, also known as the Marble Church

While filming Magnificent Monuments *Fred was working on the third-tallest steeple in England at St Walburgh's in Preston*

A church steeple has a big nut and great big long bolt that comes down the middle, as shown in Fred's drawing

It's really interesting because you can see the eight stones that take the weight of the top three-quarters of the steeple. And if you look right up to the top you can see the iron cross-tree where the great nut and bolt comes through to hold the top on the steeple.'

When he stepped out on to the ledge on the outside of the steeple he had more to admire. 'You can actually see the lovely curve of the barrel shape of the steeple,' he said. 'When you look at it from half a mile up the road it looks perfectly straight, but when you get close up to it you can see it's actually barrel-shaped, like a barrel of beer. Then there are lots of lovely pinnacles on the corner with the slender supports fretworked out, and the flying buttresses that join one to the other. Ahhhh, a lovely bit of stonework really!'

Steeplejacking Days

When I was filming with Fred he was just coming to the end of his steeplejacking career and his experience gave him a unique insight into the way that churches were built. He said:

> All my working life as a steeplejack I always liked working on church steeples and towers like this. I have just done the gold leaf and all that on two clock faces, and I am presently working on the third tallest steeple in the whole of England. It's on the church of St Walburgh in Preston and it's 311 feet high. I think I'm getting to the end of my days as a steeplejack, you know, I think it's the last big steeple I'll ever mend. When people see the enormous thickness of the walls at the base of a church tower, they have this strange idea that the steeple is almost solid, but it isn't. A steeple is very thin and very light. So why doesn't the top blow off in a gale? The answer to this is quite simple. A church steeple has a big nut and a great big long bolt that comes down on the inside for maybe 25 to 30 feet. The nut is screwed on at the bottom, and that has the effect of clamping the capstone down to the top 20 feet of masonry. That's why the

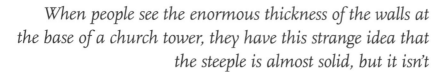

When people see the enormous thickness of the walls at the base of a church tower, they have this strange idea that the steeple is almost solid, but it isn't

weather vane stays up there in a gale, and doesn't end up down in the cemetery.

The trouble is that, as the years go by, the rain gets in at the top of the steeple and runs down the rod. Then the rod starts to get rusty. Some church steeples have access to get at this rod, so that you can paint it. The big steeple in Preston that I have got my ladders up has got this very thing. It has four cast-iron, Gothic-arched opening lights that have never opened for years, and they have that awful half-inch-thick dark-green glass in them. Part of the glass is broken, so that you can see the rod inside and you can see that it's in quite good condition. The top 30 feet of this steeple is only about 2 feet 6 inches across – really fine, just like a needle. It's so fine, in fact, that the top of the steeple rocks in the wind. Because of all the rocking, the head joints have cracked. Now, if this was a chimney stack, the obvious thing to do would be to put great big iron bands around it, like a set of corsets. There are some big steeples that have a lot of ironwork on them, but I think it would look rather unsightly on this one and I think I can get away with less drastic methods of keeping the thing up. First of all, though, I'm going to have to do an inspection; which entails dangling 300 feet up in the sky and leaning outwards on the end of a rope to look at the ornamental bits to see if there is anything ready for falling off.

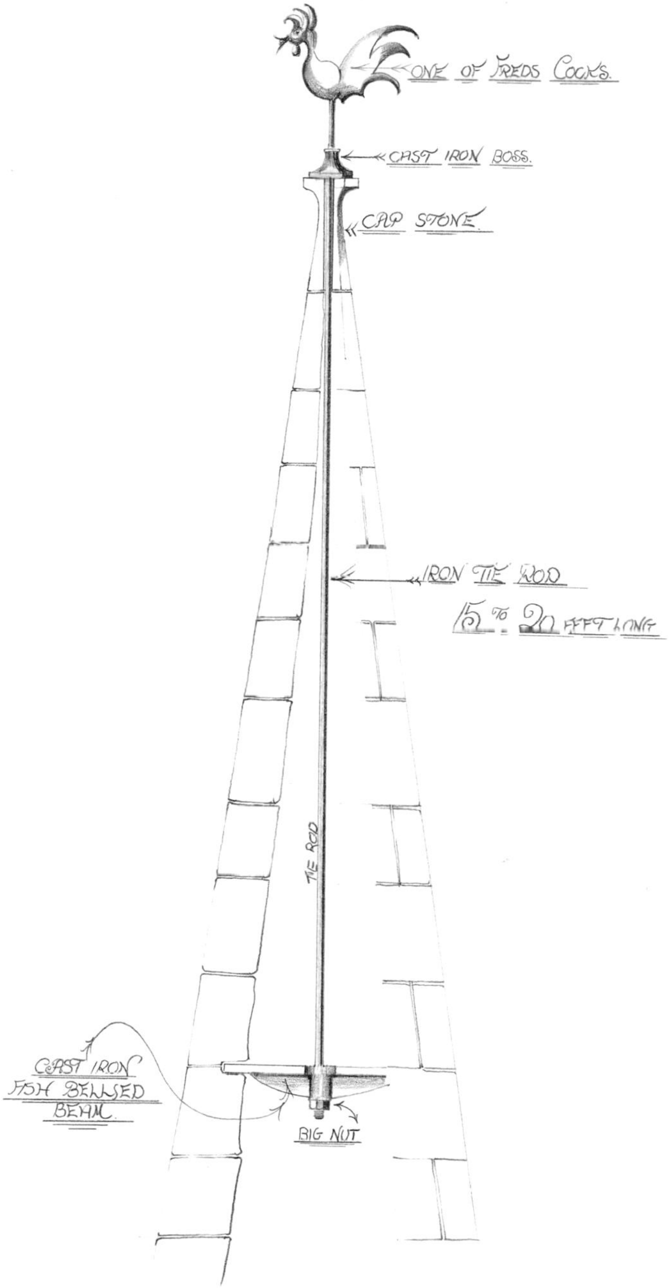

CHAPTER THREE

HOUSES AND PALACES

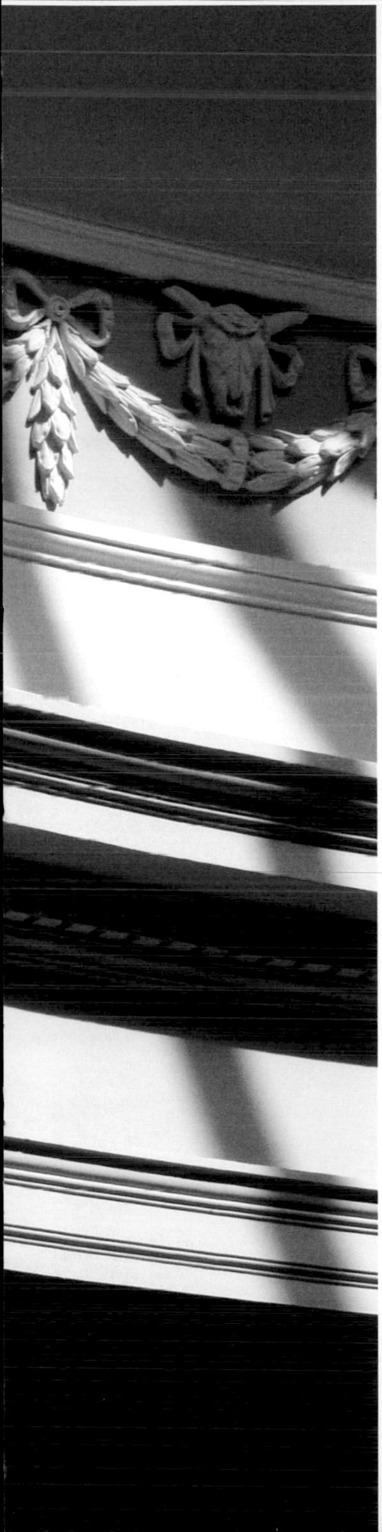

As the turbulent and violent days of the Middle Ages came to an end, the design of buildings began to change dramatically. The site of a building and its design no longer had to be chosen with defence as the main consideration. It was a time of change from fortified houses and castles to grand country houses which were much more comfortable places to live in. But the wealthy folk of Tudor England wouldn't have been able to take possession of such magnificent new homes if it hadn't been for the skills and efforts of one particular type of craftsman. It was the age of the carpenter – a time when large parts of Britain were still covered with forests and the skills of the carpenter reached their peak.

The changes that turned an Englishman's castle into his home didn't happen overnight. It was a very gradual process that began back in the medieval era. The growth of religious communities and the building of the great cathedrals had brought about the development of new construction techniques. Master carpenters began to develop specialized jointing techniques and make big advances in the mechanics of structures. By the time of the Tudors they had found ways of spanning wide spaces with massive timber roof trusses and they built huge timber-framed houses which still stand as secure and solid today as when they were first built. The carpenters who built these places were the great engineers of their day.

Medieval Manor Houses

Stokesay Castle

Stokesay Castle is in Shropshire, close to the mid-point of the English border with Wales, and in it we can see the first stages of these changes. It's England's oldest moated and fortified manor house and it has hardly altered since it was built in the late thirteenth century. It was built by a leading merchant, Laurence of Ludlow, who created a comfortable residence for himself and his family. What he wanted was a house that was good to look at, but one that had some defensive capabilities as well. The house he built took advantage of the newly established peace on the Welsh border.

The Ludlow family's money had come from the wool trade. Laurence of Ludlow built up the business which had been started by his father and became the greatest and wealthiest wool merchant in the whole of England. In the Middle Ages this was a very profitable business to be in and merchants like Laurence became the new rich of the day, with enough money to build a place like Stokesay Castle. Until this time, merchants had lived in houses in town. Only the landed gentry could have afforded something on this scale. While Laurence was having it built he obtained a licence to crenellate from Edward I. This meant that he was allowed to make his manor house

Fred tries his hand at renewing the lath and plaster of a medieval manor house

The South Tower, Stokesay Castle, Shropshire – England's oldest fortified manor house

The timber-framed gatehouse at Stokesay Castle

defendable by adding battlements. Edward had just defeated the Welsh prince Llywelyn and was building his chain of castles over on the Welsh coast. So it seemed a good idea to have some form of defence in a vulnerable border area like this, where nobody knew how long the peace would actually last.

But a fortified house was also a great status symbol designed to impress the neighbours and, although its exterior is castle-like, with walls, towers and moat, it's not really serious defensive architecture: it could hardly have resisted a determined attack. What these things did was to provide security against peacetime disorders and criminal activity – burglars or unruly mobs – which were ever-present threats. Although the Great Hall is protected by a moat, its large windows facing on to the outside world are an indication of the new sense of security being enjoyed by the end of the 1200s. The whole effect of the tower with its battlements is very impressive but it certainly wouldn't keep out an army. Up on the battlements, though, you get extensive views over the surrounding countryside so at least you'd have plenty of warning if an enemy was approaching. And it has got some defensive features like arrow loops for firing crossbows.

With jettying, the joists of a floor that is above ground level were extended outside and beyond the external wall below

On the other side of the courtyard the castle's timber-framed gatehouse wouldn't have been very secure either. It wouldn't have kept out a determined band of robbers, let alone a small army. But what we see today with all its elaborate ornamentation was only built in the seventeenth century to impress visitors. It replaced a more substantial stone gatehouse, which must have had a tower with a drawbridge across a moat which would have been filled with water. Inside the gatehouse, the courtyard is surrounded by the remains of the curtain walls which would have been much higher when the castle was first built, making this courtyard feel much more enclosed. Opposite the entrance is a single range of buildings. These form the manor house built by Laurence of Ludlow, still almost complete and unaltered since the thirteenth century. All the walls are built of local stone – mainly mudstone rubble laid in horizontal courses.

At one end of the range is the North Tower. It is one of the oldest buildings at Stokesay and because of this the ground-floor and first-floor rooms are the most castle-like. In these rooms you can see that this part was built more for defence than comfort, because its outer walls only have narrow pointed arrow loops rather than the windows you get in other parts of the manor. But it's all very different when you go upstairs. 'Up on the second floor,' Fred pointed out, 'the outer walls are all timber-framed. This must have been an important room for the family and their guests. It's got good views, but you'd feel a bit exposed if the castle was attacked and you'd only got wattle and daub between the timbers for protection. Obviously when they built it they weren't expecting any serious attacks.'

Jettying

These walls actually project out over the stone walls below. It's a building technique known as jettying and it was developed during the late medieval period. The North Tower at Stokesay has a very early example of this and Fred explained how it works:

> Joists are the beams used to support a floor. With jettying, the joists of a floor that is above ground level were extended outside and beyond the external wall below. They would hang out in space by anything from 18 inches to 4 feet. As a result the floor above was larger by that amount. So when you stand near the windows in the big room on the second floor of the tower which is jettied out there's nothing between you and the moat below other than the floorboards. In a way it's a very clever way of pinching another room above that's maybe as much as eight or nine feet bigger than the room below.

When they put the floor joists on to these, there could be as much as two to three feet of an overhang

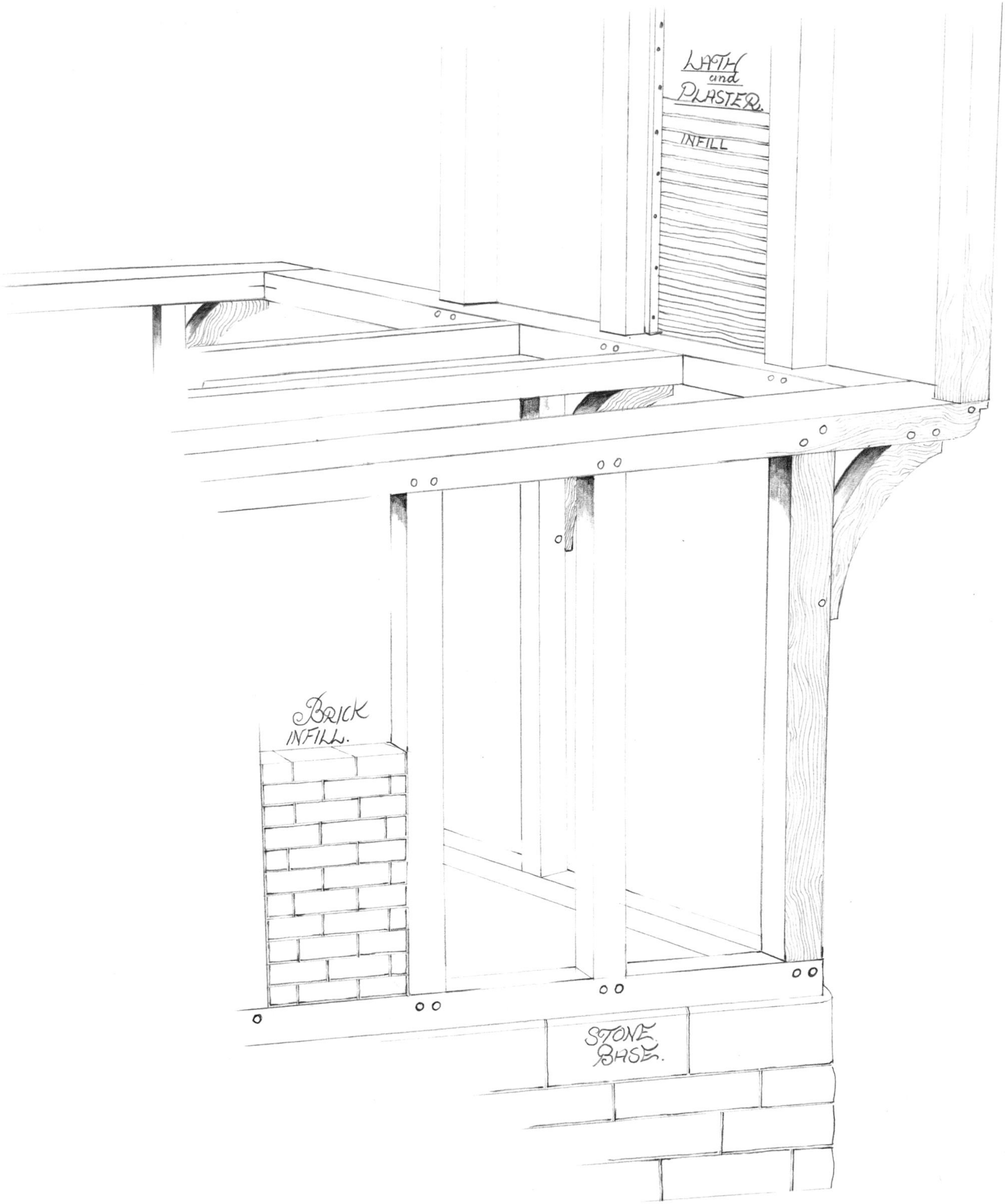

Jettied first storeys at Lacock, Wiltshire, would once have accommodated hand looms

In my drawing the vertical timbers are the outside walls of the lower chamber. When they put the floor joists on to these, there could be as much as two to three feet of an overhang, which of course in the case of both sides of a room made the room six foot bigger. And then to compensate for this overhang they made the rather nice little brackets or bracing pieces I've drawn which helped support the floor above. These are held together with mortice-and-tenon joints and pinned across with wooden pins and they rest on stone corbels on the outside of the building which give them support.

Jettying, however, wasn't just the preserve of the rich. Lacock is near Chippenham in Wiltshire and it was on the old main road from London to Bristol. In the Middle Ages cloth was our most valuable trade and the village of Lacock was an important manufacturing and trading centre. Many Lacock villagers lived by weaving and, by the fourteenth century, cloth was being made there and sold far and wide, even exported from Bristol. In the fifteenth century wide looms were introduced and many houses were built with broad first-floor rooms to accommodate them. The jettied first storeys of some of the half-timbered houses you can see in the village would once have accommodated hand looms.

In medieval times only local materials could be used to build the houses so the earliest ones were all built of timber, plaster, thatch and stone. Bricks, brick tiles and slates came much later. Although many of the cottages and houses have been altered since the Middle Ages, some of them still have features dating back that far and Lacock is as good a place as anywhere to go to get a feel of what a medieval trading centre would have been like.

Jettying, then, would become very common in timber-framed buildings of the Middle Ages, but the North Tower at Stokesay is a very early example of the technique, where they actually built the jettied wooden structure on top of an existing stone tower to create an extra floor. In terms of defence or security from bands of robbers or local insurgents,

They must have been pretty skinny in those days to get up stairs like this

though, it was the South Tower at the other end of the range of buildings that was the most secure part of the manor house. To get into it there would have been another drawbridge at first-floor level and you can see the sawn-off lifting gear still in place here. Inside there was a series of self-contained apartments that may have served as sleeping quarters for the lord and his family and for important guests. These en-suite rooms would have been the height of luxury for the day. They have latrines built into the walls and the windows have window seats. By looking at the timber beams we can see how the roofs/floors were built in the towers and the size of room that could be created without needing any internal supporting walls or columns. To get from one floor to the next there is a spiral staircase.

'They must have been pretty skinny in those days,' Fred said, 'to get up stairs like this. But how did they get their furniture up into these rooms? A room like this would have had beds, chests and chairs. So how did they get them in here?' The answer can be found in a blocked-off door now containing two windows on the second floor. A door at this height could hardly have been reached from an external staircase, so it may have been used for the lifting of furniture that was too large to be brought up the narrow spiral staircase inside the tower.

Cruck Beam Construction

Stokesay is full of timberwork on a grand scale, especially in the Great Hall. Fred described it:

> As you can imagine from its size this was the main public room of the manor house. The effect of size and space is increased by this magnificent roof, which, in its time, was leading-edge timber structural technology. The six large windows were glazed in the upper half but only shuttered in the lower half because glass was very expensive, so the shutters would have been kept closed in wet or windy weather, making candles or torches necessary in here. The hall would have been used for meals. A stone hearth in the floor in the middle of the hall marks the site of an open fire and, although you can't see it now, there would have been a hole up in the roof directly above the hearth to let the smoke out. The timber staircase at the end of the hall is one of the most remarkable survivals at Stokesay, since medieval staircases like this have usually been replaced at later times. It's made with solid timber treads cut from whole tree trunks.
>
> The staircase and gallery at the top of it are supported on large timber brackets rising from stone corbels on the wall but it is the cruck-beam roof in the Great Hall that is the

Building a cruck-beam roof at Peter McCurdy's workshop, illustrated by Fred's drawing

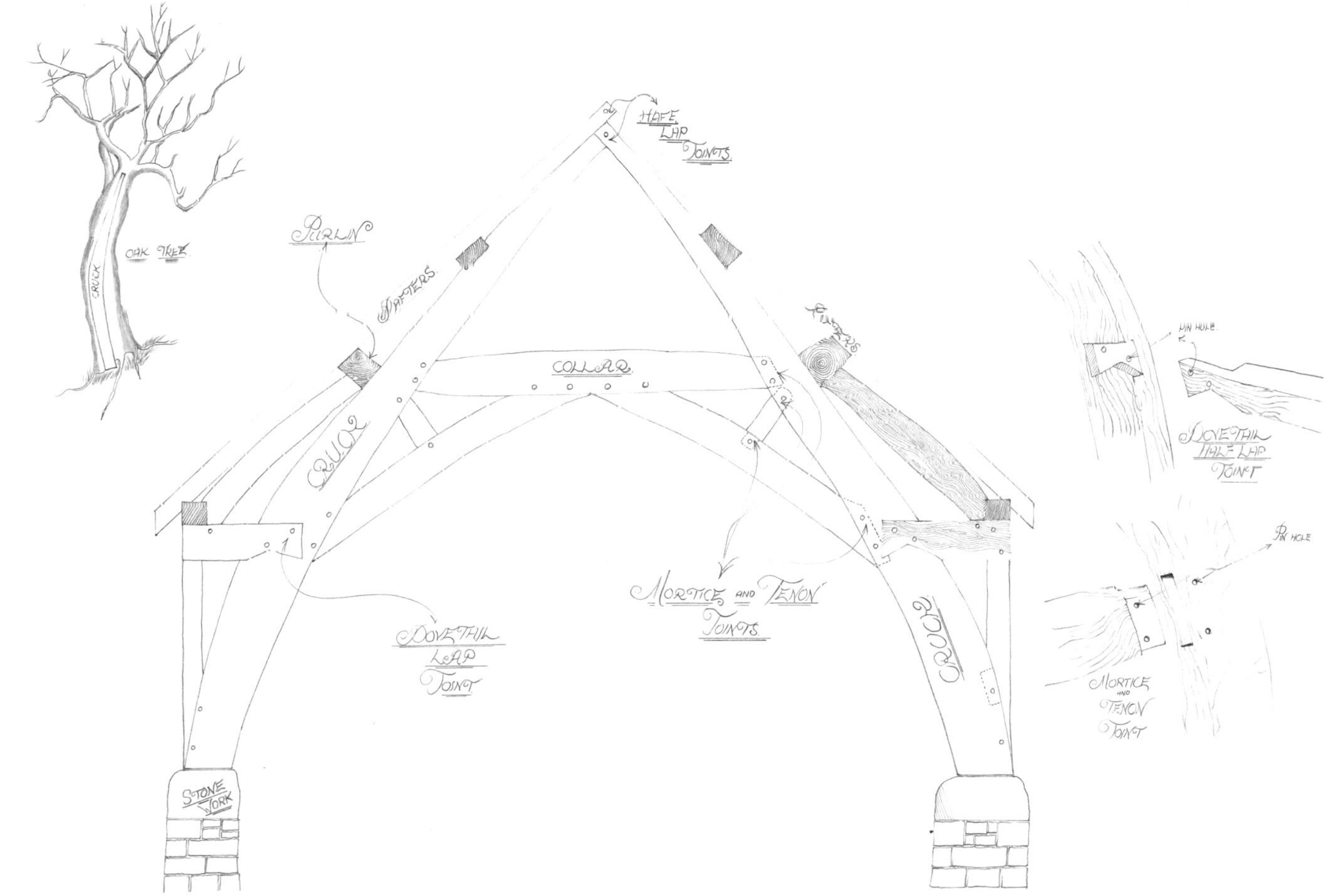

Cruck-beam construction was real leading-edge carpentry technology of the day

greatest achievement of the carpenters at Stokesay. This was real leading-edge carpentry technology of the day. Until now halls of this size had been aisled with rows of aisle posts supporting the roof, but Laurence of Ludlow wanted to clear the floorspace of his great hall under a single roofspan. This was a period of great innovation in roof carpentry and the aim was to find a way of creating single-span roofs with large timber arches across the hall.

In the late twelfth century everybody was striving to span the greatest distances with arches made of wood. The cruck beam was the answer and Stokesay has a very early example of a cruck roof. The crucks are the huge pairs of opposed, curved timbers which are resting on the stone pilasters in the walls. They form a basic frame on to which subsidiary timbers, walls and rafters are fixed. In the Great Hall three pairs of crucks rise from halfway up the wall to the top of the roof and clear the wide space with one truss. Structurally it's like a very wide arch made out of wood with the piece in the middle being held in place by compression like the keystone in a stone arch. The little drawing I've done shows the basic principles of the cruck-beam roof construction. Basically it's two bent trees and they must have had a man going round all day looking for the right trees for the job. A bent tree is roughly the shape of one of these great beams that come up from the wall at the side and basically they just leaned one of these on the other up in the middle of the roof and did a few basic joints, nearly all half-lap joints, and plenty of oak pegs to hold it all together. Then they put a horizontal collar or beam across the top to give it even more stability and stop it collapsing inwards. That would then be braced by small internal bracing pieces and various bits that were shoved in anywhere really to give it a bit more support.

Cruck construction was used in both single-storey and two-storey buildings and was in common use for better-class houses until about 1600. Both the cruck roof and the timber staircase at Stokesay indicate that Laurence of Ludlow was employing a carpenter who was at the forefront of advances

It was a bit unusual to see a seventeen-year-old lad knock his mother's chimney down and build another one

in this great period of new and experimental techniques in carpentry. Fred observed:

> Some people say craftsmanship like this has disappeared today, but it's not. The man who keeps those skills alive today is Peter McCurdy. Peter is a specialist in timber frame construction. I first met him at Shakespeare's Globe, where his firm built the timber-frame, then I went to visit his workshops deep in the Berkshire countryside where they were building a cruck-beam roof that was very similar to the one at Stokesay for a barn at Glastonbury that had been destroyed by fire. When we were filming we were able to see all the stages of building the Glastonbury barn from construction in his workshop to erection on site in Glastonbury. The process is exactly as it would have been in the Middle Ages other than the modern tools that are used now. Carpenters have always played an important part in the development of buildings. Often their work is hidden away under the work of the stone masons as it was in the great castles and cathedrals. But with timber-framed buildings and great beamed roofs it's there for all to see. It's good to see that the skills of building a cruck-beam roof that were first developed at the time that Stokesay Castle was built over seven hundred years ago are still alive today.

Fred's First Chimney

> One of my earliest adventures into building was when I still lived at home with my Mum and Dad. I lived in a row of terraced houses similar to Coronation Street and we had a chimney stack which had four pots on it. Only one of these pots was actually in use and that was the one for the back kitchen fire and the back boiler. The other three just dripped water down the walls of our bedrooms and next door's as well. So good old Fred, who was newly trained as a joiner and a roof-mender amongst other things, decided that he would take the old chimney stack down, and build a new one for his mother. So I set off with brand new Accrington bricks and there were various comments in the neighbourhood like: 'What's he doing?' because it was a bit unusual to see a seventeen-year-old lad knock his mother's chimney down and build another one. I didn't have any real design or plans for it, but as I built it, I kept getting another idea about what I wanted it to look like, and when it was completed it turned out to be a most handsome thing. It has stood there now for some forty years, and when my mother died, my brother and I sold the house. The people who live in it now decided they wanted to knock the chimney down, but the local authority put a preservation order on it. So it is still there till this day, and looking just as good as the day that I built it. I'm quite proud of it really.

They expanded the house as their needs dictated, making it more comfortable

Ightham Mote, one of the oldest and lovliest medieval manor houses in England

Ightham Mote

It was partly these experiences handling his own building materials that gave Fred such a feel for old buildings and the people who had built them:

> Ever since I was a lad, I've been interested in building and building techniques and in that great range of skills that go into building a house. They're skills that have been around for a long time – since way back into the Middle Ages and even beyond that. Ightham Mote in Kent is one of the oldest and loveliest of medieval manor houses to have survived in England. It was never a grand place and it has stood largely unchanged for over 650 years. One of the reasons for this is that its successive owners were people like squires, sheriffs and members of parliament and they were all fairly indifferent to changes in fashion so they expanded the house as their needs dictated, making it more comfortable, but retaining the basic shape and style of the buildings. When they did add to them, it was always in a manner that was sympathetic to the medieval origins of the house. The history of the house is quite complicated, but it has become possible in recent years to trace it in much greater and more reliable detail because the National Trust, in the course of repairs that have been carried out there, has had to take sections apart and in doing this they've unearthed some clues about its construction.

A moat surrounds all four wings of the house and all of its walls drop straight down to the waters of the moat. Within these four wings there is a lovely open courtyard. When you first look at it, all of the house appears to have been built at the same time, but it is actually the product of six centuries of development. The main entrance to the house is on the west front via a castellated central gatehouse tower. A manor house of reasonable size like this required some sort of formal entrance that also guarded the main approach.

From the outside the east front is the most complex of the four façades because it was constructed and reconstructed at many different periods, and the builders didn't think twice about breaking through earlier walls to light a room, or to insert a bank of chimneys, or build upwards into a third storey or outwards on to ledges which had been left vacant. All this was done in what seems a fairly haphazard way. The north front is more regular because most of it was constructed all at the same time in about 1480. The south front, built in the late fifteenth or early sixteenth century, is the most photographed because it looks the most genuine. But looks can be deceptive and its attractiveness is partly due to some twentieth-century renovation.

What made Ightham Mote particularly interesting when we went to look at it with Fred in 1999 was the restoration work being done there. It was the largest conservation project ever undertaken by the National Trust on a building of this age and right from the start the Trust decided to retain as much as it could of the original material, where it hadn't decayed beyond redemption. When they did have to provide new stone for the sections that were too badly weathered, care was taken to avoid excessive smoothness in the finished appearance and, when they were repairing the interior surface of the walls for replastering, traditional laths were used to follow the undulations of the timber frame. Where it was necessary to cut away sections of the frame because the wood had been eaten away by moisture or woodworm or dry rot, the new oak was fitted to the old by imitating medieval carpentry joints.

A house like this has stood up to the elements for centuries, so how did they manage to build things that lasted for so long? The materials they used must have been pretty good. And before they had all the things that modern builders have, they had to use whatever materials were to hand – usually from the fields around.

Fred found out about this when he had a go at mixing the plaster they were using for renewing the lath and plaster walls and he was given a bucket of cow dung to throw into the mix. He was told that it gives the plaster more elasticity and that it helps it to harden well.

Lath and plaster was fine for the house of a country squire in rural Kent, but the next place that Fred went to see was a palace built to entertain a king. Ightham Mote is quite a modest house but Hampton Court is the most impressive palace of Tudor England.

Early Tudor Houses and Palaces

Lath and plaster was fine for the house of a country squire, but not for a palace built to entertain a king

Hampton Court Palace

The first buildings at Hampton Court belonged to the Knights Hospitallers of St John of Jerusalem, but in 1514 a ninety-nine-year lease on the estate was taken by Thomas Wolsey, the Archbishop of York and Chief Minister to Henry VIII. By 1515 Wolsey had been appointed both a cardinal and Lord Chancellor of England, and he began a building programme which turned his new country seat into a home that was fit for a man of his status. It had to be built of something that looked more substantial than a house like Ightham Mote. But again it came down to the availability of local materials, so Hampton Court was built of brick.

Throughout the 1520s, Wolsey used his residence for pleasure and for affairs of state, but by 1529 he had fallen out of favour with the King because he couldn't secure the Pope's consent to Henry's divorce from Catherine of Aragon. He was forced to hand over Hampton Court to the King, who couldn't wait to get in. Within six months Henry began his own building programme. In just ten years Henry VIII spent more than £62,000 rebuilding and extending Hampton Court. This was a vast sum at the time, equivalent to approximately eighteen million pounds today.

What makes Little Moreton Hall in Cheshire so distinctive is the variety of patterns within the timber frame

The central gatehouse in the middle of the West Front is part of Wolsey's original palace, but the wings on either side of it were added by Henry VIII and once contained the Great House of Easement, or communal lavatories, and the kitchens. One of the best surviving parts of Wolsey's Hampton Court is the vast outer Base Court, the first courtyard you go into on entering the palace. The main shape of the court, which had forty guest lodgings arranged around the courtyard, still remains. Some of Wolsey's rooms, probably built in the late 1520s when he needed to retire away from the king and queen, still survive. They have been altered and restored but many original features have survived including the plain Tudor fireplaces. Much of the Tudor palace was decorated with a type of wooden panelling called linenfold, which was intended to create the effect of draped fabric on the walls. Two of the Wolsey Rooms still have this sixteenth-century panelling thought to have been produced for Wolsey because they incorporate a cross design. Fred was impressed and explained how this effect was created. 'The timber in between all the folds,' he said, 'would be done with concave and convex moulding planes and a small grooving plane. It looks to me like it could have been done the same as masonry with a hammer and chisel.'

The hammerbeam roof in Hampton Court's Great Hall

By the time Henry's work on the palace was finished around 1540, it was one of the most modern, sophisticated and magnificent palaces in England, complete with tennis courts, bowling alleys, pleasure gardens for recreation, a hunting park of more than 1,000 acres, kitchens covering 36,000 square feet, a fine chapel, a vast dining room, a great hall, and multiple *garderobes* or toilets, which could sit twenty-eight people at a time. Hampton Court remained much as it was in Henry's day until William III came to the throne in 1689. William then commissioned Sir Christopher Wren to rebuild Hampton Court. Wren's original plan was to demolish the entire Tudor palace, except the Great Hall, which he intended to keep as the centre of the new northern approach. However, neither the time nor the money was available for this ambitious plan and Wren had to be content with rebuilding the king's and queen's main apartments, on the south-east side of the palace, on the site of the old Tudor lodgings. What we see today is largely the way the palace would have looked after William's alterations and extensions had been completed.

Hammerbeam Roofs

Hampton Court is one of our grandest palaces, a place that really was fit for a king, but for Fred the crowning glory was Henry VIII's Great Hall with its splendid hammerbeam roof. It is the largest room in the palace, 106 feet long, 40 feet wide and over 60 feet high; it was begun by Henry VIII in 1532 to replace a smaller hall on the same site. The hall had two functions: first, to provide a great dining room where six hundred or so members of Henry's court could eat; and secondly to provide a magnificent entrance to the State Apartments, which lay beyond. The hammerbeam roof, designed by the King's Master Carpenter, James Nedeham, is richly decorated with royal arms and badges and a series of carved, painted heads, but it was the technology of its design that interested Fred.

> A hammerbeam roof is a roof structure that was developed to obtain a wider span by the use of cantilevered beams projecting from the wall and supporting elaborately braced trusses. I always thought it all came about because they couldn't get trees that were big enough and I'm sure basically that's right; that you're limited by the length of trunk that a tree could provide for a beam and if you imagine spanning 40 feet like they have in the Great Hall you'd need a beam of immense depth if you could find it. But there's a little bit more to it than that. Basically horizontal brackets project from the top of the wall like a cantilever at wall-

For Fred, the crowning glory of Hampton Court was Henry VIII's Great Hall with its splendid hammerbeam roof

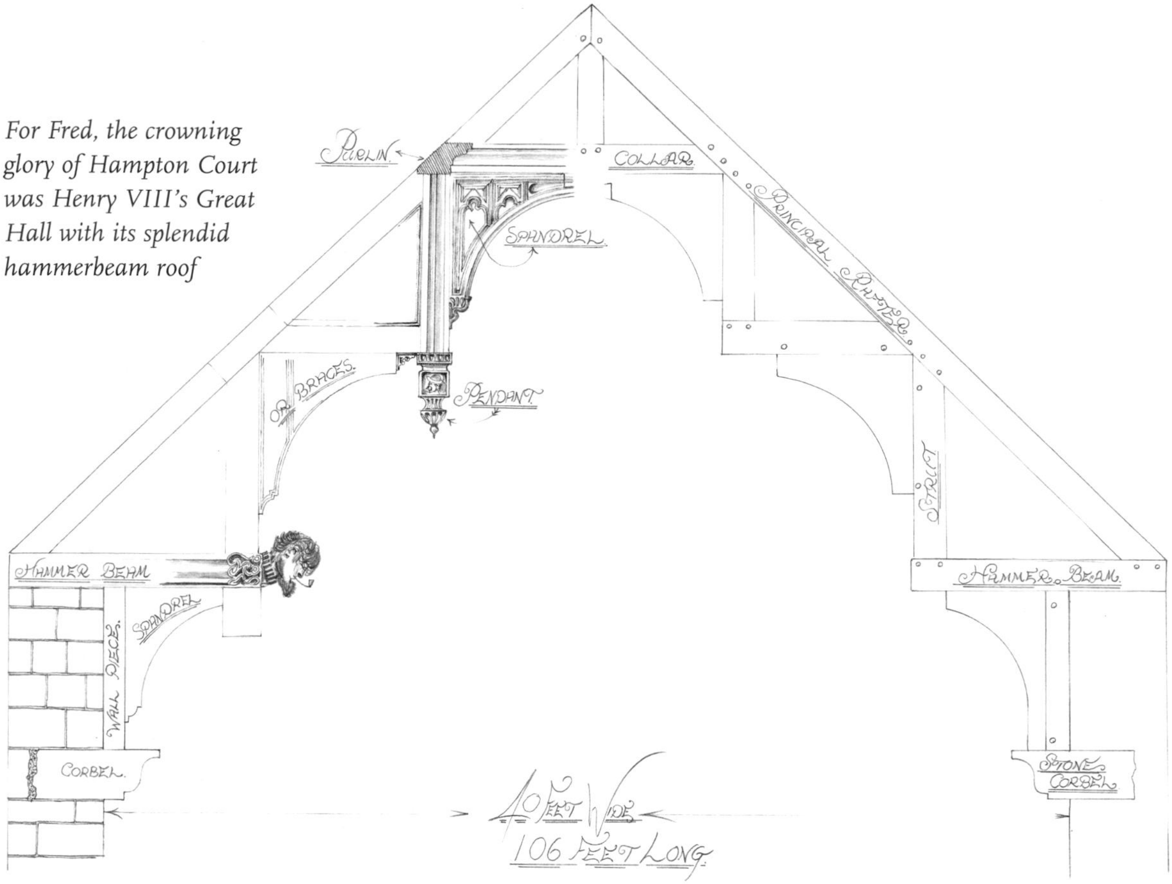

plate level. The inner ends carry vertical timbers called hammerposts and an arched brace which spans the centre of the room. On top of the brace there is a collar beam which goes across the room at a narrow point right up in the rafters. It's a very light construction that looks a bit like the underside of a ship and I think it's Henry's best bit of building here.

One of the secrets of the carpenter's trade was the variety of joints he would use on a building like this to hold it together. In the Middle Ages the main joint that was used for constructions like hammerbeam roofs and cruck-beam constructions was the mortice-and-tenon joint. Basically this is a joint formed by a rectangular projecting piece, the tenon, fitting into a rectangular socket which is the mortice. The tools needed to make such a joint are fairly simple and even today they must be very similar to those they used in the Middle Ages: a big drill for drilling a series of holes in the beam where you are making the mortice, then of course a chisel and a hammer for joining up all the round holes into the rectangular shape of the mortice. When they knocked the tenon into the mortice it would open up the tenon to get a really good grip, so that once you get it in you won't be able to get it out. That's how all the great trusses in Hampton Court would have been made. They were all clever lads with these fancy joints.

Looking out over the Clock Court – the Great Hall is on the far right

Timber-frame Construction

I've always been interested in carpentry on a large scale. I once had a friend called Ken Devine who started life off as a joiner and ended up as a fiddle player in the Hallé Orchestra. And he were a bit like me; he were a frustrated steeplejack and also had a great interest in coal mining and we promised ourselves in the pub when we'd had a pint or two that one day we would build a wooden pithead gear in my back garden. We got started on it, but poor Kenneth, alas, didn't live long enough to see it finished. Really it's not everybody in a residential area who's got a pithead gear in their back garden. But to me it's rather a handsome piece of carpentry and all built very much the same way that they would have done a big timber frame like this in the Middle Ages.

Back then it was the age of the carpenter, a time when timber-frame technology was reaching its peak and, as England settled down to a long period of internal peace, the castle walls of a house like Stokesay were no longer necessary. Fred travelled to Cheshire to see one of the finest examples of timber-framed architecture in England.

Little Moreton Hall

Little Moreton Hall near Congleton is one of the finest examples of timber-framed architecture in England. It's typical of the late medieval–early Tudor period when a moated

I've always been interested in carpentry on a large scale

house served to display the owner's standing rather than defend him. Fred explained:

The main building material of medieval houses, especially in the north, was still timber. To build a house like Little Moreton a plinth of brick or rubble was put in place first and this supported a framework of vertical posts on which horizontal beams were placed to carry the walls and roof. Posts set closely together were infilled with lath and plaster to give the characteristic black and white appearance known as half-timbering. The posts are discontinuous from one storey to another so the load-bearing is spread. Considering the amount of acreage of land that the Moretons owned, they mustn't have been short of a few oak trees when they started building this place.

The timber would still be in the round when it arrived on site and would be split with iron wedges and cut into timbers by men using pit saws. The timbers were then cleaned up and shaped with an adze. Then the mortice-and-tenon joints would be worked on to the ends of each piece of wood. The frame of the house was assembled on the ground in sections which would be raised into position and secured with wooden pegs. The upright pieces of the frame, called studs, were fitted into a horizontal sill, which rested on a stone base. I made a small model to try and portray how they went about building half-timbered houses in Tudor times. One of the first pieces to be erected on the plinth would be the corner post and then the first of the cross-members would be attached to this. All of these would have been marked so the builders would know exactly where they fitted. When you look at half-timbered buildings, you will notice that the vertical timbers are never very long, maybe just 10 or 12 feet. This meant that two or three decent lads of reasonable fitness and strength could get one of these timbers and manhandle it up and more or less shove it together very much like I did with my pithead gear.

The timbers were firmly joined using oak pegs which were first heated to make them shrink a bit. Holes just big enough for the shrunken pegs were made in the two timbers to be joined, then the peg was hammered into the holes. When the peg got damp as it picked up moisture from the air, it returned to its original size, giving a snug fit that made the joint firm. Of course the more pegs you have in the holes to hold it all together the better. Once they'd got to the top of the walls the top rail would be dragged up, no doubt on ropes tied to a couple of pieces of fair poles. To fill in all the spaces in between the framing, the first

Little Moreton Hall near Congleton, Cheshire – one of the finest examples of a timber-framed manor house

ideas they come up with were like lath and plaster which is really just chopped sticks that were nailed into a sort of rebate, and then they plastered it with a mixture of cow dung, sand and lime and this is what is referred to as wattle and daub. They did this on both sides of the wall, inside and out, so there was an air cavity in between which of course were good insulation. When you look at these buildings, though, you often see examples of great cracks all the way around the edges where the wattle and daub had shrunk and left a gap where it meets the timber frame and I suppose the draught howls in there in the winter.

Another weakness of half-timbered buildings is the bottom ends of the vertical framing because where they touch the stonework there were no fancy damp courses or owt like that so some of the timbers must have gone rotten where they sat on the damp plinth. The trouble was that some got rotten quicker than the others which is why when you look at this place

the thing goes up and down all over the place and all the horizontals end up higgledy piggledy. It almost looks as though it could all come tumbling down at any time. Wherever you look at it from, it's impossible to find a perfectly straight edge anywhere on the building. Each floor has a very pronounced dipping under the pressure from the heavy timbers and the stone slate roof and the contorted structure appears to be on the verge of collapse.

One of the things that makes Little Moreton so distinctive is the great variety of patterns within the square panels of the timber frame. The black bit of the pattern is all wood and the things that look like four-leaf clovers are quatrefoils which have been made out of solid wood in a style that is peculiar to this region. The more wood you had on show in a house like this, the wealthier you were, which made the Moreton family, who owned the hall, pretty wealthy. They had been powerful local landlords since the thirteenth century – serving as mayors of Congleton and tax-collectors for the king. The house was built by three generations of the family over a period of 120 years between the middle of the fifteenth century and the end of the sixteenth. During this time they doubled the size of their estate by buying up land that had come on the market following depopulation after the Black Death and the sale of church land following the Dissolution of the Monasteries. That's how they paid for a place like this.

As the house was developed and enlarged it formed a bit of a hotch potch of

The bay windows at Little Moreton Hall added in the mid-sixteenth century

As the house was developed and enlarged, it formed a bit of a hotchpotch of buildings

buildings around the central courtyard. The earliest parts are the East Wing and the Great Hall, where little has changed since they were modernized with the addition of gabled bay windows in the mid-sixteenth century. The carpenter who did the work was Richard Dale, as he announced with an inscription on one of the windows – *Richard Dale Carpenter made this window by the grace of God* – an early bit of advertising. We don't have any record of how much time Richard Dale spent on this work or how many men he had working for him, but it must have taken a heck of a lot of man hours to do woodwork as elaborate as this and on such a grand scale. What we do know is that he became a good friend of William Moreton and they spent a lot of time working together on their plans for the house.

As well as the elaborate carpenters' work, it's the patterned glazing of the windows which gives Little Moreton its complete sixteenth-century character. By the middle of the sixteenth century there were fifteen glass factories in England. It was expensive and could only be made in small pieces, which is why we get the distinctive leaded windows. When the glass was put into the windows no cementing was done: the lead holds it all together.

The Great Hall, entered through an ornate porch, is the focal point of the house. It was the centre of the busy Moreton household, serving as a dining hall and meeting room for the official business of the manor. Most house plans of this time were arranged around a great hall like the one at Little Moreton. It was developed from the Norman manor, but now elaborately panelled in oak with a decorative timber roof or elaborate plaster ceiling. This was the main room of the house. As in an older great hall like the one at Stokesay, heating would have come from a fire in the centre of the hall with a hole in the roof for the smoke to escape. All the fireplaces at Little Moreton were put in in the sixteenth century and the brick chimney stacks were built on to the outside of the house.

It wasn't until the 1570s, over a hundred years after the Moreton family started to build the hall, that work on what is now the front of the house began. It was about this time that long galleries had started to become fashionable. A long gallery was an extended room up at the top of a house that was used for recreation and entertainment and they were being incorporated into all the fashionable new houses that were being built at the time. William Moreton's son, John, built this part of the house and he decided he would have to have one. Fred explained what the trouble with it was:

You can see the way the weight of the Long Gallery at the top is pushing the rest of the building down

He seems to have had the idea after construction of the south range had begun, so that when he had it added on to the top – disaster! When you stand at the front of the building you can see the way the weight of the Long Gallery at the top is pushing the rest of the building down. John Moreton put it up on top of his new south range but there was no support in the roof to take the weight of it. This was after the time of Richard Dale. I'm sure if he'd been here, there's no way he would have let it go ahead without any adequate support. A timber frame is a very secure and solid form of structure unless it is messed about with, which is what has happened here. Because this gallery seems to have been a bit of an afterthought the frame is not properly braced to carry its weight. The resulting forces of compression, tension and torsion made the timbers of the frame below bend, which could have led to the collapse of this part of the building if efforts hadn't been made to prop it up.

Down below the Long Gallery in the guests' parlour you can indeed see the timber frame bending under the weight of the gallery above as well as a big supporting wooden pillar that has been put in to prop it up. But the amount of weight that a timber frame like this can support is still extraordinary.

Looking at Little Moreton, Fred concluded:

The amazing thing is that the whole weight of all the structure, the roof and all the timber framing and all the floors, is all resting on the masonry at the bottom. It's not actually fixed to it in any way. And when you look closely at it you can see that even though there is evidence of many repairs that have been done over the centuries to the timberwork at the bottom, not one iron dowel has been found. The whole thing is just resting there, like a great big dolls' house. But as a joiner what I really like about it is the way that all that lovely timberwork is all on show. But why don't we have more half-timbered buildings like this? Well, they might be good to look at but you talk to anybody who's tried to live in one through a British winter and they'll tell you that they're pretty cold and draughty.

Harvington Hall

By the end of Queen Elizabeth's reign in 1603, more new houses were being built of brick. Fred explained why:

It was warmer, drier, and altogether more comfortable to live in. So if you'd got any money and you lived in a medieval timber-framed house, you'd either build a new house or modernize your existing one by encasing

Under the fashionable brick façade of Harvington Hall there is a timber frame

the earlier timber frame in brick. Harvington Hall near Kidderminster is a good example of this. It's an Elizabethan manor house, built in the 1580s by Humphrey Pakington, who was a leading Catholic at a time when practising the Catholic faith was against the law. The house is full of surprises. On the outside it's brick – much more modern and fashionable at this time than a half-timbered place like Little Moreton. But when you get inside you find a very similar timber frame to the one at Little Moreton. So the carpenter still had a very important role to play. Although it looks as though it's all built of brick, the solid-looking brick walls on the outside are not load-bearing walls. They are just a facing and all the load-bearing is on the timber frame inside the brick. And it wasn't just the framework of the house that called for the skills of a good carpenter.

Harvington Hall contains the finest series of priest holes anywhere in the country. Priest holes were hiding places for Catholic priests which were built in the late 1500s and early 1600s when it was high treason for a Catholic priest to be in England. Four of Harvington's are sited around the great staircase and they show the trademarks of Nicholas Owen. Owen was trained as a carpenter and mason and became one of the greatest builders of such hiding places in history. The English Catholic community needed priests, but harbouring a priest was an offence that was punishable by death by public torture. Owen built dozens of hides and saved the lives of many priests. When he was caught he was taken to the Tower of London where he was tortured in a particularly gruesome manner to try to get his secrets out of him. One report said: 'They tortured him with such inhuman ferocity that his stomach burst open and his intestines gushed out.' But he never disclosed a single detail of the hiding places he spent his life building.

Altogether there are eight hides scattered around the house. They are not all Owen's as some are very easy to find. Owen's hides were much more cunningly disguised. Fred found one on a raised platform at the end of a small

At Lacock Abbey, the cloisters of a thirteenth-century nunnery were incorporated into the design of the house

room, and he showed how the panelling at one end of a cupboard on the platform could be removed to show a brick and timber wall. One heavy upright timber hangs on a pivot which, when pressed at the top, swings outward to leave a gap which a man could just squeeze through. This, he explained, was a particularly sophisticated triple hide which had three lots of things to get through in order to find it; he demonstrated how the hiding place itself is between the ceiling of a passage outside the room and the floor of a room above. This is the most ingenious in the house and one of the most cunning and complicated in the country. It was so good that it was only discovered in 1894.

Lacock Abbey

Harvington Hall was always a place to live, but some great Tudor houses didn't start off as houses at all. We went down to Lacock in Wiltshire with Fred to see another place that has religious connections. It's best known today as the home of a famous nineteenth-century inventor, because it was here that William Henry Fox Talbot perfected his calotype technique and made what is generally acknowledged to be the world's first photographic negative. Lacock is a place of pilgrimage for photographers from all over the world and a photographic museum commemorates the achievements of Fox Talbot, but the abbey is also of great architectural interest. It contains a fascinating amalgamation of three different periods of development, described by Fred:

> William Henry Fox Talbot was a great nineteenth-century innovator who were responsible for finding out more or less all we know today about photography. His family home, like a lot of big country houses, started life as a religious institution. This was acquired by a wealthy family at the time of the Dissolution of the Monasteries and turned into a Tudor mansion. Later on, in the middle of the eighteenth century, it was in the forefront of the Gothic Revival, when the lofty great hall and dining room were rebuilt in the Gothic style by John Ivory Fox Talbot, the famous photographer's great-grandfather. What you've got now then is an interesting mixture of styles – medieval cloisters, all the Tudor bits and the Gothic Revival stuff – a bit of a hotch potch really, but all very pleasing.

At Lacock Abbey we see how, in houses like this, the original features of the abbey such as the cloisters of the thirteenth-century nunnery were incorporated into the design of the house. The Tudor house has an unusual octagonal tower and courtyard of domestic buildings as well as additions from the eighteenth and nineteenth centuries including

the Gothic hall. Before the Reformation there were usually anything from fifteen to twenty-five nuns belonging to the order of Augustinian canonesses at Lacock. After Henry VIII appropriated it in 1539, he sold it to a courtier, Sir William Sharrington. Many courtiers and nobles like Sharrington bought monastic buildings that Henry had taken over and the king made rather a lot of money out of it. But having bought the monastic buildings, how was Sharrington going to be able to live there? Fred explained:

> The abbey buildings that he had bought consisted of a church and large rooms for the nuns. They were cold and uncomfortable and it certainly wasn't the kind of place where a wealthy courtier would want to live. Things were a bit rough. I mean, in the whole nunnery there were only one fireplace in a room called the warming room. This was a problem that was arising all over the country, as buyers tried to think of ways of making use of old monastic buildings that had been taken over by Henry and sold to them. The answer for most of the people who purchased these places was to flatten the whole lot, then use the materials to build a new house. It was a lot easier than digging the stuff out of a quarry. At some places, like Buckland Abbey, the church itself was converted into a house with the addition of walls, ceilings and floors.
>
> William Sharrington didn't do this. Instead he demolished the church, converted the first-floor rooms of the nunnery, then built a tower and a stable court. It was this building that put Sharrington's mark on Lacock, turning it into a Tudor building stuck on to a medieval church building. The result is a house with four sides round a central courtyard and the whole house is full of long narrow passages which follow the line of the cloisters underneath. It's an interesting amalgamation of medieval nunnery and Tudor house with some eighteenth century Gothic additions.

Eighteenth-century Elegance

Many of the great houses of the eighteenth century served as magnets for the leading architects and designers of the day. Castle Howard in Yorkshire, designed by John Vanbrugh in association with Nicholas Hawksmoor and built in the first decades of the eighteenth century, was the first major expression of the Baroque in English domestic architecture. The dramatic and highly ornate Baroque style in art and architecture had developed from around 1600 in Europe, particularly in Italy and Spain. When it spread to England, it found expression in monumental and highly ornamental buildings set in grand landscapes. Vanbrugh and Hawksmoor went on to design and build another great Baroque house, Blenheim Palace. It is regarded as the highest and fullest expression of English Baroque, but it was not to everyone's taste. To some eighteenth-century patrons and their architects, the style seemed heavy, indulgent and over-elaborate. They looked back to more sober classical buildings designed a century earlier in a style that was derived from the work of the eminent sixteenth-century Italian architect Andrea Palladio. Its essence was a revival of classical Roman symmetrical planning and harmonious proportions which aimed to recapture the splendour of antiquity. Its first exponent in Britain was Inigo Jones, one of the greatest English architects. With

Fred has a go at making the sort of ornamental plasterwork that was a distinguishing feature of Robert Adam's work

With its great classical columns, Kedleston Hall, Derbyshire, epitomizes the fascination the ancient world held for eighteenth-century architects

the building of the Banqueting House in Whitehall and the Queen's House in Greenwich, Jones became the pioneer in England of a classical style of building inspired by both ancient Roman and Italian Renaissance architecture. Jones's elegant buildings were a major influence on future generations of patrons and architects. The Palladian movement resulted in the building of more restrained houses on a villa plan.

The House of Dun

In the eighteenth century the leading Scottish architect of the day, William Adam, started to design houses that broke away radically from the Baronial style seen at its best in Glamis Castle. Fred described the House of Dun enthusiastically:

> Like Glamis they started off with a great tower here. But you won't find any traces of it in the house that stands on the site now because here, rather than building round an existing tower, they demolished it and built a completely new house on a greenfield site just down the hill from where the tower had stood.
>
> Work began on it in 1730 and it took over ten years, a bit like one of my jobs! But really you can see why if you compare it with Glamis Castle, with its rough stone and the big wide joints. The stonework here is beautifully finished; the joints in the stones are so fine, you cannot even get your fingernail in. Every stone's done individually and everything fits together to such a degree of perfection that if you put your eye to the corner, all the stonework in the wall is dead straight. You cannot fault it; in fact it's straighter than what they would get it these days, I reckon. Outside the front door they have beautiful reeded columns, but the great glory of the interior of the House of Dun is the Saloon. This magnificent plasterwork by Joseph Enzer was the final touch when it was being built. Enzer received a payment of £216 not just for the plasterwork in the Saloon, but for the whole house. Most people coming into the Saloon wouldn't have a great deal of idea how all of this magnificent work were done, but going back to my days at art school, they had an ornamental plastering department and even though I never did any myself, I always took great interest in what were going on in there. And, of course, they made nearly everything on flat benches and then glued and screwed them to the walls.

Robert Adam

William Adam had set a new trend for house-building in Scotland. But it was Adam's more famous son, Robert, who took some elements

Kedleston Hall: one of the earliest and greatest of Robert Adam's works

of this style and added a lot of ideas of his own to create a style of architecture that is named after him. Robert Adam had spent three years travelling around Europe drawing and studying great buildings from the past. He was particularly impressed by the remains of the ancient Roman buildings he saw and it was this that influenced the Adam style more than anything else. He adapted the richness of Roman design and ornamentation he'd found in his study of classical remains to create his own distinctive style. When he returned to England in 1758 he began to work in a broadly Italian style and within the space of five years he had effected a revolution in the design of interiors. Many of his commissions were fashionable remodellings of existing country houses. Into them he brought classical columns and porticos and created spatial drama and complexity by juxtaposing rooms of contrasting shapes and sizes with columned screens and semi-domes inspired by the baths of ancient Rome. He revived the Roman technique of stucco and used this for all kinds of ornamentation as well as for whole ceilings. He liked to introduce curves into his designs and used alcoves and apses often with screens of columns.

One of the principal elements of the classical architecture that inspired Adam was the column with base, shaft, capital and

Fred's drawing of a Corinthian column

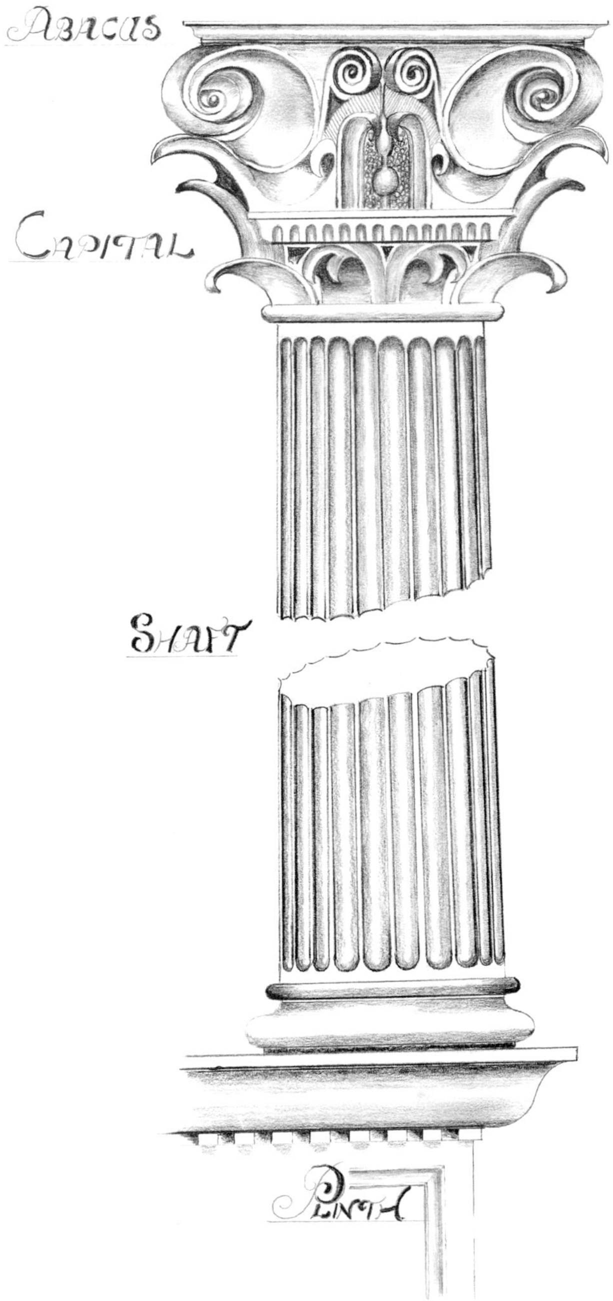

entablature. Columns were decorated and proportioned according to one of the accepted styles or orders: Tuscan, Doric, Ionic, Corinthian or Composite. The simplest of these is the Tuscan, which was derived from the Etruscan-type temples of pre-Roman central Italy. Doric is even earlier in origin and is divided into Greek Doric and Roman Doric; the former has no base but both have a fluted shaft with plain rounded capital. The Ionic order originated in Asia Minor. It has a fluted shaft, more slender than the Doric, with base, and a capital with curving ornamentation based on the shape of a ram's horns. The Corinthian order, with fluted shaft, base and capital with ornamentation based on acanthus leaves, was an Athenian invention, but was later developed by the Romans. The Composite order is a late Roman combination of elements from the Ionic and Corinthian. Both Greeks and Romans used these styles in a definite sequence, with Doric or Tuscan on the ground floor of a building, Ionic on the first floor and Corinthian on the second.

Kedleston Hall in Derbyshire is one of the earliest and greatest of Adam's works. With its great columns and magnificent marble hall it sums up the fascination that the worlds of ancient Greece and Rome held not just for Adam, but for the eighteenth-century landowners like Kedleston's owner, Lord

Corinthian columns at Kedleston Hall

Scarsdale, who commissioned him. The main central block was intended both to entertain important guests and to be an elegant setting for Lord Scarsdale's collection of pictures and sculptures, and Adam's Great Hall, 70 feet long and 40 feet high with Corinthian columns supporting the coved ceiling, was a true temple of the arts. The south front that Adam created is based on the triumphal arch of Constantine in Rome. Behind it is the domed Rotunda, even higher than the Great Hall that it adjoins.

Adam's success at establishing his own fashionable and personal style in the 1760s led to an immense output from the architectural office he ran in partnership with his brother James. His style rapidly matured and the interiors of Harewood House, Syon House, Osterley Park and Kenwood are masterpieces of design. Everything is characterized by the meticulous attention to detail of a man who was determined to set his own mark on everything from the finish of the painted decorations and stucco work right down to the carpets, furniture and lock-plates.

Culzean Castle

Culzean Castle on the Ayrshire coast is one of Adam's most important and distinctive works. Here he was commissioned to rebuild an existing castle and his plan was to transform it into a romantic-looking house that made the most of its dramatic clifftop setting. Adam worked on Culzean over a period of fifteen years from 1777 to 1792.

First he incorporated the original building, which had been a stronghold of the family for centuries, into the south side of the mansion. He squared up the original tower house and added a three-storey wing on either side. The sandstone was quarried locally, some actually coming from beneath the castle itself when it was removed to make the foundations and the cellar.

Culzean Castle, Ayrshire – the north wing with its drum tower and the grand dual staircase

The next stage of his building work was the north wing on the seaward side. Here he built a big drum tower with rooms on either side of it to sit right on the edge of the cliff. It's a sheer drop for about a hundred feet down to the shoreline. Immediately below the tower there's a great cave and, said Fred, 'he must have been a bit unsure of himself because he built a big stone pillar in the middle of the cave just in case. But he obviously built the tower for the beautiful panoramic views of the countryside, of the sea and everything. It's quite a magnificent thing really, perched here right on the edge. It's stood the test of time as well because it's all still here. It's slightly eroded because it's facing the western elements and the Atlantic, so it's taken a bit of a beating over the years it's been here.

Adam's brief extended to the whole of the Culzean estate. 'Not only did he build the house,' Fred said, 'he built a wonderful viaduct that's part of the grand entrance to the place. The building that houses the clock tower on the other side of the courtyard was already there but you can see how he added to it, smartening it up with towers and turrets and mock battlements on the top and a new skin down the front. And the castle's farm that was built to his design is a work of art in itself. You won't find many farms that look as good as this.' Fred then turned his attention to the stonework:

> There's a massive amount of stonework here, a lot of it with quite delicate detailing. The only trouble is that all the stone came

I had the pleasure of meeting the resident stonemason and having a go with the hammer and chisel myself

from a quarry that the earl owned. It's all sandstone, not the hardest-wearing stone, and it's all very exposed to the elements up on this Scottish cliff edge, so over the two hundred years since all this was built it's got very badly eroded. They've got a resident stonemason here on site who is kept busy all the time restoring and renewing some of the stone that has suffered most and I had the pleasure of meeting him and having a go with the hammer and chisel myself.

But Adam's work isn't just about stonework and exteriors. He designed his buildings to achieve the most harmonious relation between the exterior, the interior and the furniture. Especially important was his conviction that the interior of a building and its furnishings and decoration was the proper concern of the architect. The Old Eating Room with its delicate plasterwork is very typical of the airy, light, elegant style that bears his name. It was essentially a decorative style and it is as a designer of interiors like this that he is chiefly remembered. He gave meticulous attention to every part of each room, from the carpets to the most unobtrusive decoration so it is one of the outstanding features of an Adam interior that everything, even the smallest detail, was part of the unified scheme created by the architect.

Right at the heart of Culzean is one of Robert Adam's crowning glories: the grand Oval Staircase. When he had finished his work on the north and the south sides of the house he was left with a small, damp, dark central courtyard which still separated the two parts of the house. So, ten years after he first began work there, he used this dark, sunless space to produce a design which gives a great feeling of light and airiness. There wasn't enough room for a conventional circular spiral staircase so Adam made it oval. By placing slimmer Ionic columns over larger more decorative Corinthian columns, he created an illusion of soaring height within the stairwell. Light streams in from the cupola above and passes through the columns to create contrasts of light and shade. The Oval Staircase at Culzean is Adam's final masterpiece. It unifies the whole castle and makes the staircase appear to be the central feature rather than a later addition. The whole effect is very dramatic and it's very typical of Robert Adam – the only man in the story of the building of Britain to have a whole style of building named after him. 'The whole place,' said Fred, 'is a magnificent monument not just to the workmanship and hard graft of the men who were responsible for the building of Britain, but also to their imagination and ingenuity.'

Nineteenth-century Style and Revivals

Fred with James Harvey-Bathurst of Eastnor Castle and his traction engine, Atlas

By the start of the nineteenth century Britain was beginning to enjoy the fruits that came from being the birthplace of the Industrial Revolution and the wealth and prosperity this brought has left us with some magnificent buildings. It was a time when the Gothic Revival came into its own. This distinctive style was first seen in the work that Sanderson Miller did at Lacock Abbey (see page 128) and in the second half of the eighteenth century a style of romantic Gothic or castellated design began to emerge. In the following century it became one of the principal forces in shaping the face of Victorian Britain. Gothic Revival was inspired by a group of men who looked at medieval architecture and married the style of the buildings of the period with the modern techniques of their day to come up with something utterly new and amazing.

Eastnor Castle

Eastnor Castle in Herefordshire is a particularly fine example of the style and it was one that Fred was very familiar with:

> Some years ago I met the gentleman who owns a particularly fine traction engine called *Atlas*. When I came to visit him it turned out that

Built in the nineteenth-century, Eastnor Castle married the style of medieval architecture with modern construction techniques

Using the technology of the Industrial Revolution, the architect Robert Smirke created a country house that looked like a medieval castle

he lived in a castle. At least it's called a castle but it's a bit of a fake really. It's never been under attack or under siege or anything like that because, as castles go, it's all rather new. Eastnor Castle was only built in the nineteenth century and it's actually a great country house that was built to look like a castle. Its owner is the Honourable James Hervey-Bathurst who took a bit of a shine to me because of my interest in steam engines. He's a collector of traction engines himself, a great railway enthusiast and an admirer of things of an antique nature in general.

We got quite chummy actually and I ended up helping him repair and restore one of his best engines: *Atlas*. This lovely old engine once lived in the north of England but it now resides on the forecourt of his castle in Herefordshire. That's really how we became friends. In the castle there are great piles of staircases wherever you go and, as a stranger, it's easy to get lost. I would imagine living in a castle like this would have been a bit lonely at night. It's so far away from any street lighting that it's a bit dark, a bit like being down a coal mine.

It's so far away from any street lighting that it's a bit dark, a bit like being down a coal mine

Eastnor Castle was built for the second Baron, later first Earl Somers, by a young architect called Robert Smirke between 1810 and 1824. It was a time when the size and splendour of a country house were the most obvious indications of the standing and fortune of the family who lived there. Earl Somers wanted to show the world how rich and important his family was; and in those days, a big country house was like having a big car today: it showed everyone just how grand and important you were. He wanted a house that would speak of authority in a time when the aristocracy felt very threatened by the radical fervour of the French Revolution and the Napoleonic Wars. So Smirke persuaded him to adopt the new Norman Revivalist style that was coming into fashion. Eastnor would hark back to the great age of the English kings, when baronial power and royal authority were unquestioned. It would send out an unmistakable message to the people around it. It said: 'Remember who your masters are.'

Smirke had already built a couple of buildings in this style, but Eastnor was to be his unquestioned masterpiece. Not only would it look like a medieval castle, it would be built on a similar grand scale. It took fourteen years to build, and its final cost was £85,923 13*s* 11$^{1}/2$*d*. The stone for the site was brought by canal all the way from sandstone quarries in the Forest of Dean. In the first year alone, 250 men put up 4,000 tons of building stone, using 16,000 tons of mortar and 600 tons of wood to construct the walls. Like Harlech Castle, these walls were built directly on to the granite bedrock, with hardly any need for foundations. Fred explained:

> Smirke was able to build his castle bigger and better than any of the castles of Edward I because he had something which Edward's master architect did not. Up among the roof trusses of Eastnor, you can see that the main structural beams of the castle are made of cast iron: a relatively new technical innovation of the time. These give the castle much greater strength than the old timber trusses which the medieval builders had to rely on, and are the precursors of the reinforced concrete frames with which we build today. The thing about using cast iron was that it was a lot lighter than wooden beams so it was a lot easier to get the roof up here. It was a lot cheaper than wood and you could span a wider space with cast iron. It enabled Smirke to design a massive Norman-style hall with minimal use of wood, which was in great demand for naval building because of the Napoleonic Wars.

Using his drawing Fred was able to elucidate further:

The reason for the great cast-iron beam at Eastnor Castle is to actually hold up the front wall of the tower. This wall is situated roughly halfway along the Great Hall, so instead of building a great big arch like the Normans would have done, they cast two iron beams. These are held together by big bolts with big spear-headed nuts. The whole roof is made of iron, there's no wood at all. The only wood involved is the actual ceiling of the Great Hall. When you get up inside the roof space you find that all the metalwork is prefabricated units. They're all pieces around 8 feet long with dovetails on the end and mortice holes and pin holes and holes for keys to be knocked through. This meant that when they got them up here the rate of assembly would be very quick and of course there's no dry rot in it or woodworm or any of the problems you get with wood. It'll all be there in a thousand years if you keep going up and giving it a coat of tar every now and again.

Mr Smirke made very clever use of cast iron in the building of the castle, not only for the structural part of it, but the ornamental bits as well. If you study the staircase, at first it appears to be made of wood, but the posts are cast iron and the way that they would do this was to make first of all a wooden pattern and bury that in the sand in a moulding box and then pour in the molten iron. The great beam that we looked at up in the rafters would be made in exactly the same way, but on a much mightier scale. It wasn't the only modern technique that Smirke used. At the time that Eastnor was being built steam engines were being used to power various sorts of machinery and no doubt they were used here for stone saws and big wood saws and for lifting the building materials. A steam engine meant that great amounts of stone could be cut and made nice and smooth in a fraction of the time that it would have taken to do the job by hand. One man could now do what hundreds were needed for in the past and they were able to more or less mass produce beautiful detail and ornamentation.

The Great Hall, Eastnor Castle – Fred's drawing shows how, above the ceiling, a cast-iron beam supports the weight of a tower

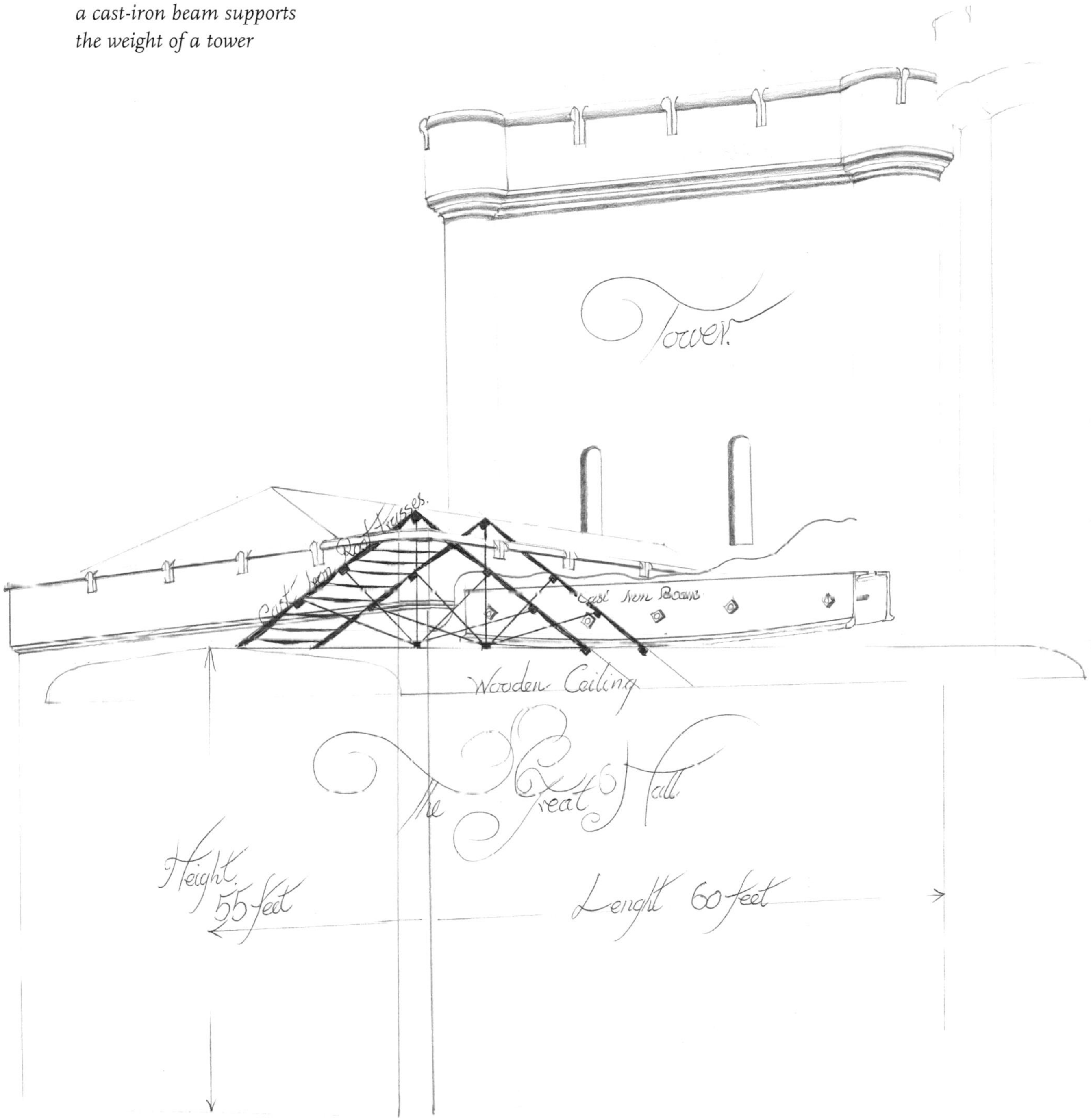

The results of the use of the new technology of the Industrial Revolution are stunning. As you go into the castle, sweeping between two suits of armour that flank its entrance, you are first met by the Great Hall, a vast space with a massive decorated arched door at the far end, reminiscent of a great Norman castle. This is the symbolic heart of the castle. Originally, the hall was completely studded with arms and armour, deliberately emphasizing its medieval nature. But a few years ago (in 1989), most of the armoury was moved to another of Smirke's interiors, the Red Hall. As you come through from the Great Hall into the Grand Staircase, you step from the rounded arches of the Norman style into the pointed arches of Medieval Gothic. In fact, there is a whole mix of styles on the staircase, both ancient and modern. Most of the decoration is Italian Renaissance, with the wooden chandelier actually coming from the Palazzo Corsini in Florence. But the banisters are made of good old English cast iron, and fashioned in a classic Medieval Gothic style.

You enter another period as you step into the Dining Room. Now, you are in a great panelled Tudor Hall, reminiscent of the manor houses of the Middle Ages. Smirke originally designed this to look like an early medieval hall, with Gothic arches over the doors, but these were later covered over to give the effect we see today. All the furniture in the room was designed by Smirke, though the original table surface has been replaced.

The most magnificent room in Eastnor, however, was not designed by Smirke. The Gothic Drawing Room was designed for the second Earl Somers in 1849 by Augustus Welby Northmore Pugin, and it is a magnificent example of the fully developed Victorian Gothic style which Pugin

The Gothic Drawing Room designed by A.W.N. Pugin

The Victorians loved to have everything beautifully ornate, very pleasing to the eye

pioneered. The whole room is modelled around the Somers family crest. At the centre of the room stands an ornate Gothic fireplace sporting the family crest, above which Pugin has painted the Somers family tree snaking up to the ceiling. The work was done by John Crace and his brother, who did all of Pugin's most important work. It caused such a stir when it was being done that the locals believed Queen Victoria was coming to stay. 'For me,' said Fred, 'this magnificent drawing room is the height of Victorian splendour and embellishment. It is a very fine example of how good they were at decorating places back in them days.'

Fred had a great love not just for the Victorian era and the heroic age of engineering that came with it, but also the rich decoration of their buildings. It was the age he loved; the age he would have liked to have lived in.

> The Victorians loved to have everything beautifully ornate, very pleasing to the eye. The ornamentation were almost as important as the building itself and it's always been my favourite style. I first became interested in buildings when I were about fifteen years old and I lived in a small terraced house and me mother and father wanted me to be an undertaker. Now, I didn't fancy that, so I got on my bicycle and I pedalled off to the youth employment bureau where they fixed me up with a job as a joiner. My work as a joiner got me into some of the splendid mansions that the cotton mill owners and the bleach-works owners had built back in Victorian days when Britain ruled the world. The thing is, I couldn't help but notice, having come from a house that hadn't got any skirting boards, the quality of the woodwork in all these places I went into – all the beautiful skirting boards and marble fireplaces and architraves and panel doors and, best of all, the fancy plastered ceilings. It made me wonder, however did they do it?
>
> The Victorians went to great lengths to make things pleasing to the eye. Whether it were a great civil engineering project or something as small as a window catch, there was so much lovely detail and everything was much more ornate than what we do these days. It's this that I like about the Victorian era. That's why I've got so much of this sort of stuff in my house. It was a time when machinery and new materials enabled the Victorians to imitate what they admired from other periods. They made town halls look like Greek and Roman temples; they made mill chimneys look like medieval towers; they made houses look like castles; they disguised cast iron as Gothic woodwork and they built a new Houses of Parliament

The 'palace of the modern magician' – Cragside in Northumberland – home of the first Lord Armstrong

> that matched perfectly the medieval splendours of Westminster Abbey right next door. It was an age when they went to great lengths to make things good to look at; an age when the embellishment and decoration of a building was just as important as the way it was built.

The Industrial Revolution brought a great surge in house-building and the rich industrialists built wonderful mansions for themselves. From Eastnor Fred went all the way up to Northumberland to see a fine example built by one of Victorian Britain's mightiest industrialists.

Cragside

Cragside in Northumberland was the home of the first Lord Armstrong: innovator, inventor, engineer and gun-maker. In the nineteenth century he played a major role in the industrialization of Tyneside and his Elswick works in Newcastle was the heart of a heavy-engineering industrial empire.

Built on a bare and rugged hillside Cragside was one of the most remarkable houses of its day. It had hot and cold running water, telephones, a fire alarm, a hydraulic lift and – the most magnificent thing – all the electricity was generated by a hydroelectric power station. No wonder they called it 'the palace of the modern magician'. The house began as a comparatively small one. It was built between 1863 and 1866 for Sir William Armstrong, as he then was, as a weekend retreat to get him away from his armaments and engineering business in Newcastle. The architect of the original house is unknown, but in 1869, Armstrong called in Norman Shaw to turn Cragside into a proper country mansion. At first Armstrong only wanted Shaw to add to the house's existing north end, not to improve or replace it, but Shaw gave himself a bit more scope to use his own ideas and, by the gradual additions that he made over the next fifteen years, the house came to assume the wild, picturesque outline that it has today. Armstrong had a very hands-on approach and the work was done not by a building contractor but by local masons working under his direct supervision.

But it is as much for the technology as its architecture that Cragside is distinguished. Many of Armstrong's installations survive and they can be seen on the 'Power Circuit', a circular walk through the house and grounds. As well as building the house, Armstrong created a series of lakes in the grounds to store the water that provided the power to generate electricity and drive all the hydraulic machinery he installed in the house. Fred explained:

Lord Armstrong did all he could to help his domestic staff with all his hydraulic machinery. He had a lift for taking up the coal to the bedrooms, a hydraulic lift. And a wonderful spit in the kitchen driven by a water turbine quite a way off down in the cellar. And it works by a complicated system of rods and bevel gears and joints and you can actually move it away from the fire or move it into the fire, and of course it goes round.

As well as using hydraulic machinery to help with the domestic chores, he also used it to impress prospective customers. The whole place were really a shop window for the inventions that he did. The dining room is where he entertained such guests as the King of Siam and the Shah of Persia, who came here for arms dealing – to buy guns off him – and without a doubt it must be one of the finest Victorian domestic interiors in Britain. If you look at the ceiling alone you can see there's a few good oak trees gone into it and the fireplace is a wonderful creation. It's got to be the biggest inglenook fireplace in England. The outer arch is a great Gothic arch and it's survived very well, but I think Sir William did a bit of overstoking because you can see there's a few nasty cracks in his mantelpiece proper. You can imagine him sat there on a cold and frosty night thinking of what he was going to do next with his hydraulics. But one of the things that is most interesting to me about this place is that a great industrialist like Armstrong, who was responsible for many major technological advances, chose this very traditional Old English style of building for a house that he filled with modern inventions.

CHAPTER FOUR

MACHINES THAT CHANGED THE WORLD

CASTLES, CATHEDRALS, GREAT HOUSES AND PALACES all played a major part in the story of the building of Britain. But buildings were only part of the picture. In the eighteenth and nineteenth centuries Britain led the world in making and inventing things and the machines that were built here changed the face not just of Britain, but of the whole world. The textile industry was mechanized by James Hargreaves's Spinning Jenny; Richard Arkwright devised a hydraulic spinning frame in 1769; and in the same year James Watt patented his steam engine. Britain was the birthplace of the Industrial Revolution and throughout the eighteenth, nineteenth and a great deal of the twentieth centuries the country led the world in harnessing the power of coal, water and steam to drive the heavy machinery that made mass production possible.

Fred always had a passion for Britain's industrial past and its mechanical relics. In the programmes we made he introduced the great inventors from the age of steam, described the day-to-day operation of steam-powered spinning mills, rolling mills, pumping stations, railways and coal mines and painted a vivid picture of what life was like for mill hands and colliers, steel-workers and engine-drivers who laboured in industrial Britain: the workshop of the world. Fred had an almost encyclopaedic knowledge of all things mechanical and of anything connected to our industrial past. His enthusiasm for the subject was infectious and he was able to talk about his subject not just with authority but with a passion that made anybody listening to him interested.

Steam Power

Whoever turned up at Fred's house to peer through the fence at the one-man industrial heritage centre that he had created would be welcomed into the garden by Fred and given a guided tour with a detailed explanation of all the machinery he had there. At the heart of the garden was the steam engine that drove all the line shafting that powered the machinery. Engines like this had always fascinated Fred:

Raising steam – the passion of a lifetime

> There was one works near where I lived in Bolton that ran on steam. It was called T.T. Crook Stanley Ironworks and basically it was a row of houses that no doubt Grandfather Crook had bought. The building is still there to this day, but when I was a lad, it had a most unbelievable chimney stack. It was about 3 feet square and about 60 feet high and it had a lovely bend on it. They just used to burn coal dust there on a Cornish boiler which was in the back yard of the row of houses. When you looked in you could see two brick walls and in between them was the front plate of an old boiler with a water gauge and a pressure gauge on it. Crossing over the front of this there was a wooden staircase going up to a hen cabin on the top of these two walls. And when you went up the wooden staircase and through the door, there were four big iron

Replica of Stephenson's Rocket *at the National Railway Museum, York*

ROCKET.

Fred's back yard – all powered by steam

girders which went from one end of the room to the other. The girders had a steam engine on top of them with a 7- or 8-foot flywheel, a single cylinder and an exhaust pipe through the roof to the sky, so I could always tell when it was working. From this engine room a series of belts went all over the place – some spread out on this first-floor level and some went downstairs into the ground floor – a bit like my garden now really. As a small boy, I never went any further than gently creeping up the stairs and looking inside through the door and through the window. The guy who fired the boiler and looked after the engine was as black as the miners were when they came up from the pit at the end of their shift. Directly facing the premises was a pub and the engineman was always standing on the pub doorstep. You never saw him near the boiler; he spent all day in the pub. I never forgot that place. There used to be loads of engines like this where I come from – every coal mine and every spinning mill had one. But alas they've all gone now.

A steam engine really is a fascinating thing. When it's actually running it sort of comes alive in a strange way. It has an

It has been said that if you could put it in a bottle and cork it up you could sell it; it smells that good

unbelievable smell, for a start, about it. There was an old guy who came into my back yard the other day, eighty-odd years old, and he was sniffing away and he said that brings back memories from my youth: the smell of oil and steam is like a smell all of its own. It has been said if you could put it in a bottle and cork it up you could sell it; it smells that good.

Steam power was the driving force of the Industrial Revolution. It developed from the need to pump water out of mines to enable miners to dig deeper. Pumping technology was well established by the end of the seventeenth century. Simple machines powered by men, horses or water were used but by the eighteenth century they were unable to meet the demand for more raw materials like coal and tin as populations grew and towns expanded. Mines are wet places and water flowing into them as they went deeper was a major problem so some new form of pumping power was needed to remove it. Steam was the answer. The first steam engines were developed to do no more than pump water out of the mines but in time they were adapted to do much more than this. The steam engine could be used to turn a wheel at the pithead and that could be used to wind miners and materials up and down a shaft. Then as the eighteenth century progressed one invention followed another allowing manufacturers to increase their output and make Britain prosperous on a scale that no other country could match. The steam engine took over the role of the water wheel to provide the power for the new machinery of the factory age. By the nineteenth century steam power was being adapted to provide revolutionary new means of transport and Britain saw the development of the first steam locomotives and the world's first railways.

The principles of steam power are based on two major properties: first, the expansion of steam with enough force to move a piston in an enclosed cylinder; second, the sudden condensation of steam to create a vacuum and reduce the amount of force necessary to move the piston back to its starting place.

The first successful working engine was invented by Thomas Newcomen in 1712, and James Watt made big improvements later in the eighteenth century. But it had been known for a long time that steam could be used to move a mass, and many inventors and scientists, from the ancient world, through the Middle Ages, into the industrial age and beyond, turned their hand to using steam for various applications. Even today steam power is still with us as nearly all the electricity we use in the UK comes from generators driven by steam turbines.

Fred's drawing showing the workings of Hero's Temple doors as described in Spiritalia

Hero of Alexandria

The very first form of steam engine that we have any records of was invented by a Greek mathematician called Hero (or Heron) of Alexandria as early as the first century AD. Hero's interests were engineering and mechanics. No one is sure when exactly he lived, but historians have deduced from his writings that he was active in the first century AD. Hero wrote many books about ingenious inventions from gadgets to magical tricks, which included describing the first idea for a steam engine including multiple pulleys, cogwheels and levers.

Hero made a device called an 'aeolipyle' which was used as a toy. This was what Fred described as a 'steam-driven whirligig that worked a little like a modern jet engine. It was a hollow ball supported on two brackets on the lid of a basin of boiling water. The steam escaped from two bent pipes on the top, which created a force to make it spin around. The movement of the ball was used to make puppets dance. It was only in the light of what came eighteen centuries later that Hero's aeolipyle was recognized as a simple form of a steam turbine.'

Hero also described and sketched a method of opening temple doors by the action of fire on an altar. This ingenious device, which he wrote about in his book *Spiritalia*, contained many of the elements of the modern steam engine. Fred made a drawing of it:

> My drawing shows what it was and how it worked. Basically how it worked was that beneath the temple doors was a spherical vessel containing water. A pipe connected the upper part of the sphere with the hollow and air-tight shell of the altar above. When a fire was lit on the altar, the heated air would expand, passing through the pipe which would drive the water in the vessel into the bucket. The weight of the bucket then turned a number of barrels, which in turn would raise a counterbalance on the end of the ropes and open the doors of the temple. When the fire was put out the air condensed, the water returned through the siphon from the bucket to the sphere, the counterbalance fell and the doors closed.

Experiments like this with steam in the ancient world were done purely to achieve 'surprising results'. There was no need or wish to use the power of steam for any material benefits. It was for magic and religion, not industry and the economy that experiments were carried out. Ingenuity was prized more than material gain or practicality. But despite amazing spectacles like these, it was going to be another fifteen-hundred years before anybody

carried out any serious investigations into the application of steam power.

These experiments were all closely bound up with the discovery that the earth's atmosphere exerts a pressure. Investigations in the seventeenth century proved that a vacuum could be created by the condensation of steam inside a closed vessel. A Frenchman, Denys Papin, experimented by boiling water in a cylinder fitted with a piston. When it cooled, the steam that had been created by boiling the water condensed and created a vacuum. Atmospheric pressure pushed the piston down and raised a weight attached to the piston by a chain and pulleys. But Papin never developed his ideas beyond the laboratory.

Pumping Engines

The development of the world's first successful steam engine took place in what today would seem a most unlikely place. Fred explained:

The earliest successful steam engines were developed for pumping water from mines

> When you think of Cornwall, you think of holidays on its scenic coastline. But for centuries it were the world's leading place for mining tin and copper. As the demand for tin and copper grew this meant the miners had got to go further and further down, which of course left them with an unbelievable problem ... water. The problem of underground seepage plagued the management and miners alike, cutting into profits, stopping production and claiming lives, especially when the shafts were sunk on the edge of the cliffs near the coast and the workings went out under the water more than a mile like the one that can still be seen at Botallack.

The search for a more efficient and reliable source of power for draining mines was the most pressing technological problem of the time. And so the steam engine was born. By the beginning of the eighteenth century every element of the modern type of steam engine had been separately invented and applied. The pressure of

One of James Watt's earliest engines, Old Bess, *was filmed with Fred at the Science Museum*

HOMPSON ENGINEER 1791

A full-size working replica of Thomas Newcomen's first successful atmospheric engine at the Black Country Living Museum

gases had been understood, the nature of a vacuum and how to obtain it by displacing air with steam was known. All that remained was for an engineer to combine all the known facts and vital principles into a practical machine capable of economically using the power of steam. Thomas Savery, from Devon, was the first to combine all the experiments that had gone before in a machine that worked. Steam was admitted to a cylinder and then condensed by a spray of cold water. A vacuum was formed and atmospheric pressure pushed the water up through a suction pipe. The problem with his engine was that at that time the boilers could not be made strong enough to withstand the pressures needed to provide a useful forcing lift.

Thomas Newcomen

The great step forward that was needed was made by Thomas Newcomen. He was descended from an aristocratic family and had an ironmonger's business in Dartmouth. Little is known about his personality other than the fact that he was humble and was not looked upon as an individual of importance, even in his own community. In the 1680s he had formed a partnership with John Calley, a local plumber, and together they would make the rounds of the mines of Devon and Cornwall supplying their owners with metal and doing work for them on site. Here he was able to see the problem of flooding at first hand and observe the difficulty of pumping water out of the mines using horse-powered and water-powered pumps. He decided he would come up with a better method. And when he did he set in motion one of the most crucial developments of the Industrial Revolution.

Newcomen's first successful engine was installed at a colliery in Staffordshire and it proved to be the world's first commercially successful steam engine. It was used at Dudley Castle for pumping water out of the coal mines on Lord Dudley's estates. There are actually very few Newcomen pumping engines left but at the Black Country Living Museum they've built a full-size working replica of the 1712 engine with a beautiful engine house.

The 'Fire Engine', as it was known, is housed in a brick building from which a wooden beam projects through one wall. Rods hang from the outer end of the beam and operate the pumps at the bottom of a mine shaft and raise the water to the surface. The engine itself is a very simple affair, with only a boiler, a cylinder, pistons and operating valves. Fred's drawing shows how the engine works. A coal fire heats the water in the boiler, which is little more than a covered pan, and the steam that is

generated then passes through a valve into the brass cylinder above the boiler. The steam in the cylinder is condensed by injecting cold water into it and the vacuum that is created underneath the piston pulls the inner end of the beam down and causes the pumps to move. Fred explained more about its workings:

> When it's in steam, it gives you chance to go back to the very beginning of the steam revolution. When you look at it now you can see the great beams sticking out of the end of the engine house which in turn work the pump rod down the shaft. And of course that's attached to the pumps at the bottom of the sump. It's totally different to a modern steam engine. When Newcomen conceived of this engine there was no boiler technology. The only thing there was, was like a giant kettle from the brewing industry. And that's literally what the boiler is. The original had a copper bottom and a lead top, which occasionally would melt, and a cylinder mounted directly above that with a valve. Turn the steam valve off and inject water. Cold water condenses the steam and the cycle begins. Really even though it looks a bit technical it's quite simple.

When Fred was given a demonstration of the engine it was pointed out that even though it is simple it's a very difficult engine to keep running. Most of the work is in keeping the fire right because there are no other controls – no valves or anything. But as long as you keep the fire going, once you've started it the engine will continue to function

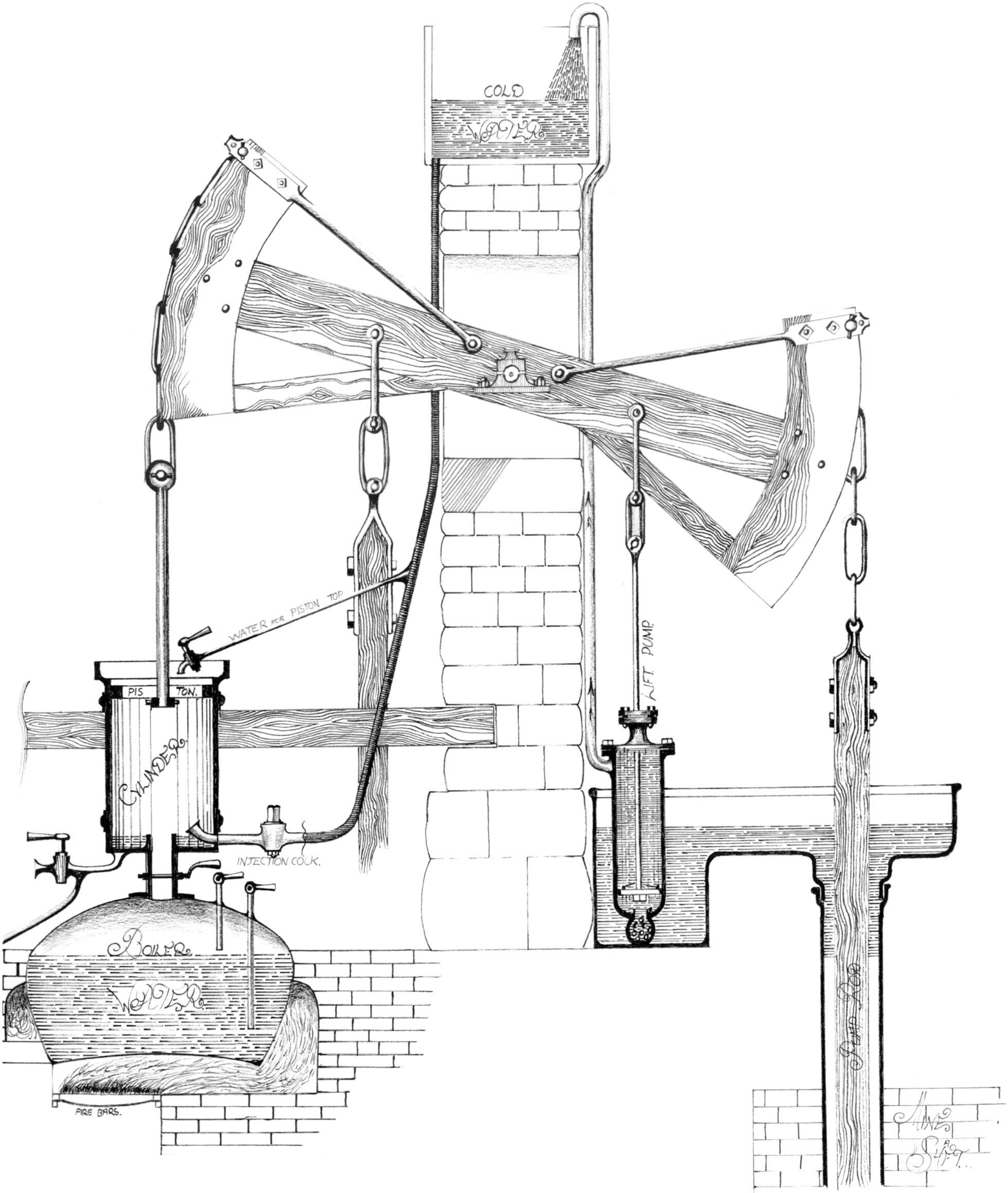
COLD
WATER
WATER FOR PISTON TOP
PIS TON
CYLINDER
INJECTION COCK.
LIFT PUMP
BOILER
WATER
FIRE BARS.

Fred's drawing shows the beam sticking out of the end of the engine house attached to the pump rod that goes down the shaft

on its own, so this was in fact the world's first self-operating machine. 'Really, in 1712,' Fred said, 'this were the cutting edge of technology. Believe it or not, this engine really was a breakthrough. The only other way they had of raising water from mine workings were by bucket propelled by horse gins. So really this were quite something when it came along. It enabled miners to go much deeper to get rid of the water.'

Newcomen's engine was an instant success. Within a few years of its invention the engines had been introduced into nearly all the large mines in Great Britain. When it was found that the new machine could be relied upon to raise the large quantities of water to be handled, many new mines, which could not have been worked at all previously, were opened. Newcomen's engine now meant that mines could be sunk to more than twice the depth that had until then been possible; within ten years it was in operation all over Europe as well.

In spite of Newcomen's unbelievable success, however, and worldwide acclaim for these engines, they had many weak points. First of all each spray of water to condense the steam cooled the cylinder. This meant you used up the next bit of steam just reheating the cylinder – not moving it. So you needed a lot of fuel to keep the steam up. Some of the Newcomen engines would gobble up 12 tons of coal a day. Now, that was bearable if you had the engine installed on top of a coal mine, but for the tin mines of Cornwall, far from the coalfields, it was incredibly expensive to use the engines. They were also pretty dangerous. The 'haystack' boilers were very primitive; they used plates of iron with the gaps packed with iron filings and horse dung – anything they could do to get it to seal – so the boilers wouldn't take much pressure and could easily explode. And the engine only produced up–down movement, which limited the jobs it could do to pumping alone. A more efficient engine was needed. It came when a young Scotsman working on a model Newcomen engine at Glasgow University realized that he could make a vast improvement on the Newcomen engine. His name was James Watt.

James Watt

James Watt was born in Greenock, Scotland, in 1736. His father was a shipwright and supplier of nautical instruments, many of which he made himself. As a child he was sickly, his health was delicate and he was unable to attend school regularly or to apply himself to study or play. He received his early education from his parents, respectable and intelligent people, and it was in his father's workshop that he acquired the skills of a master craftsman.

Botallack Mine – the firm of Boulton & Watt soon built up a large trade with the Cornish mines

He got a job at Glasgow University where he was given a Newcomen engine to repair and he realized that it was extremely inefficient. As he examined it he began to experiment and he made and patented a number of improvements. Watt found that at boiling point his steam, condensing, was capable of heating six times its weight of water. He realized that greater care would have to be taken to economize the steam so he set about making careful calculations how to minimize the loss of heat. The breakthrough came to him as he was taking a walk on a fine Sunday afternoon. He was thinking about the engine and realized that as steam was an elastic body, it would rush into a vacuum, and if communication were made between cylinder and exhausted vessel it could condense there without cooling the cylinder. This train of thought led him to devise the separate condenser – possibly the most vital improvement made to the Newcomen engine. It meant that the main cylinder could be kept hot, by drawing off steam into a separate chamber, enabling the engine to be run continually at a high temperature without cooling off and condensing the steam. The result was a machine that was immensely more powerful than the Newcomen engine which was essentially little more than a giant pump, and a saving of about 75 per cent in the amount of coal that was needed. The inventor applied for and was granted a patent in 1769. Other improvements Watt made include inventing a governor to regulate the speed of the engine; parallel motion to transfer power from the cylinder to a beam (instead of Newcomen's chain); and a design for the expansive working of steam, where the steam is admitted for only part of the stroke of the piston, and the expansion of the steam does the rest of the work.

Contrary to popular belief, Watt was not actually the inventor of the steam engine. He was, however, the man who made it vastly more efficient, and perhaps deserves his title as the 'Father of Steam'. Watt alone had turned a steam pump into the machine that drove the Industrial Revolution. As early as 1787 its adaptability for a variety of uses was described as 'almost incredible', while the *European Magazine* stated that it would 'change the appearance of the civilized world'. And it did. It led to the foundation of factories for the manufacture of textiles, iron, steel, machinery, tools and household goods in cities, changing for ever the worker's way of life.

One of Watt's main problems in his early days was that, although he had the ideas, he had no money to spend on experiments or, just as importantly, on marketing his engines. In fact he fell into debt straight away

attempting to market them and his financial straits were made even more perilous by endless patent litigation that seemed to plague every British inventor. By the mid-1760s, Watt was close to bankruptcy, so he had to find investment capital to start manufacturing steam engines. The two men who provided it, and made possible the successful development of his steam engine, were, first of all, John Roebuck and then Matthew Boulton.

John Roebuck was born in Sheffield in 1718 and became a wealthy physician before deciding to establish large-scale iron-manufacturing in Scotland. The site he chose was on the banks of the river Carron, in Stirlingshire, where there was plenty of water power, ready transport by sea, and good supplies of iron ore, limestone and coal. The first furnace was blown at Carron in 1760; and the first steam engine applied to working the blowing machinery of a blast furnace was erected at the Carron Iron Works in 1765. The pits that Roebuck owned to provide the fuel for his iron-making were at Kinneil near Falkirk and he soon discovered to his cost that he had to ensure they were kept clear of water. In order to achieve this he needed powerful pumping engines. In 1769 James

The beam of an 1812 Boulton & Watt engine stands next to an 1845 Harvey's of Hayle at Crofton Pumping Station near Marlborough

Watt was engaged by Roebuck to construct a new engine for him from designs that he had already started to work on. Roebuck took on a thousand-pound debt that Watt had built up while developing his engine and put in further capital and funding in return for two-thirds of Watt's patent. Watt's engine, however, was not yet ready and available for pumping the colliery, and Roebuck began to find himself in financial difficulties. Therefore he entered into an agreement with Matthew Boulton, in which Boulton would take a share of the business in return for paying half the expenses.

As the demand for tin and copper grew, this meant the miners had to go further down

Matthew Boulton was the son of a Birmingham silver-stamper and -piercer. He succeeded to his father's business and built up a great and very profitable establishment. He wasn't simply a businessman though, he was also a learned scholar; Watt praised him as a man of great ingenuity and foresight. By 1773, Roebuck's affairs were showing no signs of improvement, so Boulton acquired his two-thirds of the engine patent in exchange for the remission of a debt of £630 and a further payment of £1,000. Ironically Roebuck's financial difficulties were further compounded by the fact that none of his creditors 'valued the engine at a farthing'!

With Roebuck off the scene the partnership of Boulton & Watt was born. Boulton had recently set up a new factory, the Soho Works, in Birmingham, for the manufacture of silver and plated ware. For its time, the factory was highly mechanized and it could be developed for the manufacture of steam engines. Watt left Scotland for England, and the engine, which he had designed for the Kinneil coal mines, was erected at Boulton's Soho works. By the end of 1774 the engine was running and consuming only a quarter of the coal required by a similar Newcomen-type engine, or 'common engine' as Watt called it. It pumped the water that was used to provide the power for the works. In the early days of steam this was to become a very common arrangement, with steam and water being used together to run mills, factories and early engineering works.

Boulton's partnership with Watt was to be one of the most important enterprises of the century; a pioneer engineering firm which for twenty-five years was the sole source of steam power in Great Britain. It was in fact the union of one the most inventive brains of the age with one of the first great commercial intelligences, for the purpose of selling one of the most valuable things in existence: power! Boulton, like Roebuck before him, was also to face a financial crisis. Although the engine business was doing well, Boulton had speculated in other interests indirectly linked with engines. Over-production at a time of depression in 1787 resulted in the insolvency of several firms. Boulton badly needed an extension of credit. He appealed to Watt, but Watt, with characteristic caution, had already safely invested his money, and refused.

Boulton & Watt

In Cornwall the mine owners were desperately in need of pumping engines with lower fuel consumption. Cornwall had no coal of its own. It all had to be brought in from outside, so Watt's new pumping engine was very attractive to them and the firm of

Crofton Pumping Station near Marlborough, Wiltshire, and one of its Lancashire boilers

Boulton & Watt soon built up a large trade with the Cornish mines. As well as selling engines to the mine owners, the partners earned a lucrative income by charging a fee based on the fuel saved by using their engine rather than a Newcomen-type engine. Soon there wasn't a Newcomen engine left in Cornwall and the county's most famous landmark began to appear: the engine house. Many of them can still be seen today. They've survived because they were an important structural part of the engine itself. The giant beams were balanced on one of the end walls. Half was inside the engine house; the other half was attached to the pump rods leading down the shaft outside the house.

One of Watt's earliest engines can be seen in the Science Museum. *Old Bess* is a single-acting pumping engine. Built in 1777, it was the second engine to be built at the Soho Works, but it was the first to try the principle of the expansive working of steam. Watt's original system was to allow steam at boiler pressure to follow the piston throughout its stroke, but he realized that if an admission valve was shut partway through the stroke, and steam allowed to expand for the rest of the stroke, the steam could be used more economically, although the power output of the engine was reduced. With expansive working the action of the engine was irregular and hard to control and became known as 'Beelzebub'. *Old Bess* was built as a returning engine, to lift water from the mill race of the waterwheel and return it to the launder. It was a common engine for obtaining rotative motion where a natural stream didn't exist.

Crofton Pumping Station

Beam engines were not just used for pumping water out of mines. The oldest working beam engine in the world still in its original building and still doing its original job is the 1812 Boulton & Watt engine at the Crofton Pumping Station, near Marlborough in Wiltshire. Its companion is a Harvey's of Hayle beam engine from 1845. Both are steam-driven, from a coal-fired boiler, and are used to pump water into the Kennet and Avon Canal. The canal, which links London with Bristol, reaches its highest point near Marlborough in Wiltshire. Its summit level is higher than any natural source of supply water so it has to have water pumped up to it to fill the locks at the top every time a boat crosses the summit. The beam engines at Crofton were installed to ensure that there was always a supply of water to fill the locks. The two locks on the summit are 14 feet wide and over 75 feet long, and require 70,000 gallons to fill them! When steaming, the engines raise the water vertically over 40 feet to fill the locks.

Today Crofton is a wonderful working monument to the age of steam power. Its appeal lies in the fact that, when steaming, the engines perform the task for which they were originally installed: raising water vertically over 40 feet. The building which houses the engines is on three floors. At the top is the beam gallery where you can see the engines themselves. They are supported on the beam wall, which carries all the working loads of the engine into the foundation of the building. The beams are connected at one end to the pump rods, and at the other to the piston rods. On the middle floor you really get an idea of the immense power generated by the engines, as the piston rod is driven downwards on the power stroke, pulling the beam with it. On the ground floor you can see the condensers where the exhaust steam is condensed to produce the vacuum under the piston which drives the engine. You can also see the cylinder room down here; this is where the engine is actually driven from and where the steam level is monitored and regulated. Also on this level is the boiler room, where the water is boiled to produce the necessary steam. If you want to see the power of steam at close quarters this is the place to go.

Rotary Power

The early steam engines, including Watt's, could do nothing but pump water. With changing economic demands in the 1790s, largely due to the introduction of new machinery into the textiles industry, a change in the type of engines that powered this machinery was necessary too. From the middle of the eighteenth century the factory system of manufacturing had been growing rapidly and what the new machinery needed was rotary motion. At first factories had been located beside rivers, which were used to drive waterwheels – a tried and tested source of rotary power. But as time went on and the Industrial Revolution began to gather pace, more factories were built than the available water could supply. Attention turned to the new steam engines as a source of power.

With the growth of the factory system, attention turned to steam engines as a source of power

Boulton was an entrepreneurial businessman and, seeing that there would be ready sales for an engine to drive rotating machinery, he pressed Watt to develop one. Watt had to convert the oscillating motion of the steam engine beam to rotative action and the breakthrough came in the 1780s. Watt made an engine double-acting by applying steam alternately below and above the piston to produce a power stroke in both directions; simultaneously a number of other engineers introduced their own rotative engines. One of these was James Pickard, who adopted a

Matthew Bolton saw there would be a market for an engine to drive rotating machinery

Boulton & Watt's rotative steam engine was the first source of power that could be applied to drive any type of machine

crank and flywheel in his engine. This forced Watt to think along different lines and in 1781 he introduced what he described as his 'sun and planet gear'. Basically this was a pair of flywheels: the sun wheel that was fixed to the flywheel shaft and the planet wheel fixed to the connecting rod which was hung from the beam. The rocking of the beam turned the planet wheel which then turned the sun wheel which then rotated a shaft. It was this that led to his big breakthrough when three years later he came up with his idea of 'parallel motion' – an ingenious device for transferring power from the rocking movement of the engine beam to the vertical motion of the pump rod. The device was necessary because a piston can only operate on a vertical axis, while the motion of the beam runs along a curve. By 1784, therefore, Watt was able to supply engines capable of 'giving motion to the wheels of industry' and parallel motion became a standard feature on all of his engines. He was later to say: 'I am more proud of parallel motion than of any other mechanical invention that I have made.'

The Boulton & Watt engine had been fully developed by 1787. It satisfied a large part of the increasing demand for power until 1800 when Watt's patent for the separate condenser expired. During this period the firm supplied nearly three hundred rotative engines. The rotative steam engine was the first source of power which could be applied on a large scale to drive any type of machine, anywhere, at any time, needing only to be supplied with fuel and cold water. It set industry free from reliance on animal, wind or water power, thus making industrial expansion possible.

In spite of these great advances industry did not change overnight and by the year 1800 the steam engine had not made such a transformation in industry as is often imagined. Water power was still very much in evidence, but the new industries – cotton manufacture, iron, engineering and mining – were gradually changing over to the new source of power. All the new firms that were setting up were free to select their own sites. They no longer had to site their mill or factory by the side of a river, instead they went to the parts of the country where fuel for their steam engines was readily available: the coalfields.

By the early part of the nineteenth century, back in Cornwall, the increased efficiency of the Boulton & Watt pumping engines soon ensured that there wasn't a single Newcomen atmospheric engine left in the mining areas. 'But it did another wonderful thing,' Fred said:

> If you took off the pump rod and put a connecting rod and a crank on to it you could make it into a rotary engine. With this you could wind the miners down and get them back a lot faster and of course bring

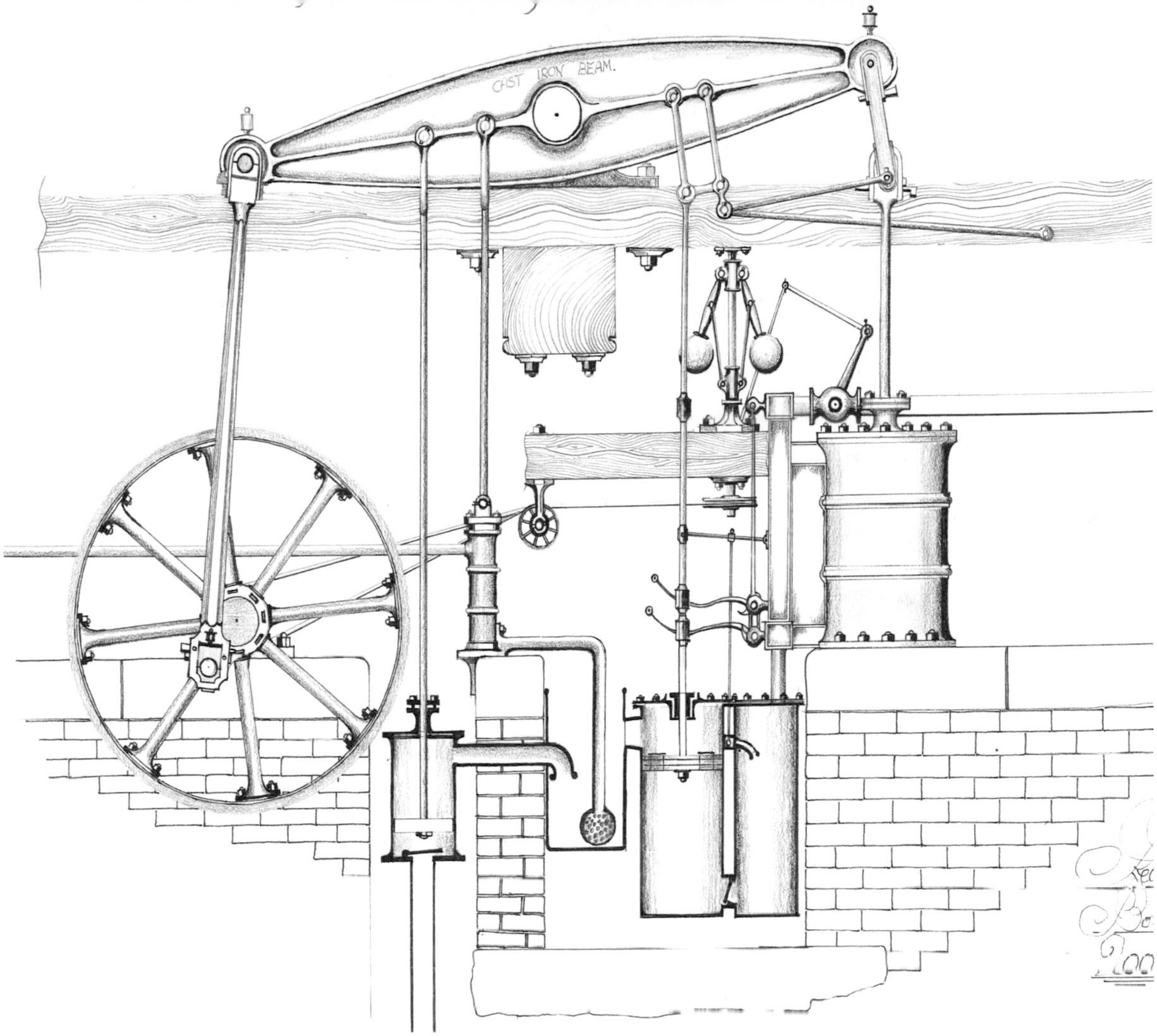

up the ore as well. This was particularly good news for the miners themselves. Many of the Cornish tin mines like Levant and Botallack were on the coast and in the olden days before the days of steam winding and wire ropes and cages the miners had to get down the face of the cliff so they were as near to the sea as they could possibly get. Then they had to enter the mine by an adit in the cliff face that met the main shaft and then continue the journey for hundreds of feet on ladders with various platforms down the shaft. Once they were at the bottom of the shaft they had to walk often for a mile or more under the ocean before they actually started work. They must have been some special men them men. But the steam winder changed all that.

James Watt might be regarded as the father of the steam engine, thanks to the patent he held with Boulton that effectively prevented anybody else doing anything. When it came to an end, Cornish engineers leapt at the chance to try out some of their own ideas. At their head was Richard Trevithick.

High-pressure Steam

Richard Trevithick

Higher press meant higher speed and paved the way for a transport revolution

Richard Trevithick invented a whole new generation of high-pressure engines with steam produced by one of his creations: the Cornish boiler. Trevithick brought about huge improvements to the old Watt engines through his amazing grasp of the principles and use of high-pressure steam – what he called 'strong steam'. Until well into the nineteenth century, many steam engines were still based on the design principles of those of Boulton & Watt – basically a beam engine with a vertical cylinder under one end of the beam and a crank under the other. They worked in the same way as low-pressure condensing engines, powered by a vacuum which was created by condensation of low-pressure steam, but from around the 1790s these engines were starting to be overtaken by new designs. In the 1790s it was Trevithick who was the first person to understand the importance of high-pressure steam. Watt had always refused to use high-pressure steam in his engines probably because of the danger involved, but Trevithick was a wilder man than Watt. He worked on the principle that higher pressure meant higher speed, lighter parts and a wider expansion stroke. He labelled his principle of high-pressure steam as 'strong steam'. His engines became known as

At the bottom of every chimney, there's a steam engine – India Mill, Darwen, Lancashire

'non-condensing' engines because they didn't use the principle of condensing steam to create a vacuum into which the piston would move; instead Trevithick used the power derived from the tension of high-pressure steam.

Trevithick was born in 1771 at Illogan near Camborne where his father was the manager of the Wheal Chance copper mine. Trevithick spent his childhood there and went to the village school where the headmaster's description of him was that he was inattentive and very slow. Fred laughed:

Bit like me in a way. He didn't do well at all and even his own father said he were a loafer. But he spent his time wandering round looking at the tin mines and the machinery that existed at the time and he amazed his superiors and so-called men of better education by his unbelievable ability for solving mechanical problems. Just by his own intuition.

By 1790, at the ripe old age of nineteen years, he'd already procured quite a few jobs as an engineer at various pits. Then his father apprenticed him to Watt's assistant Murdoch, who at that time were erecting all

Fred visited Cornish Engines at Pool to see the last in the line of Cornish pumping engines

While he was working at Ding Dong mine, Trevithick developed his first high-pressure steam engine

the great pumping engines round the tin mines. And of course you've got to rather think that Murdoch taught him all he knew. That gave him a good grounding for the beginning of his great career as an engineer and he was soon entrusted with the erection and management of large pumping engines himself. Trevithick was a skilful mechanic, but he was just as famous for quickness at calculations and for his great strength.

Trevithick constructed his first high-pressure engine for winding ore between 1796 and 1798 and so successful was his boiler development for the Cornish beam engine that the annual output of tin mined in the county rose from 2,500 tons in 1750 to 14,000 in the heyday of the industry in around 1860. The large wrought-iron boiler he designed, with its single internal flue, became known throughout the world as the Cornish boiler. His greatest advance was to design engines that would work at a much higher pressure than Watt's. If you had a hundred pounds' pressure per square inch pushing on a piston, rather than about fifteen which the earlier engines had, it was going to make the engine much more powerful and efficient. 'While he was working at the wonderfully named Ding Dong mine in Penzance,' Fred said, 'he developed his first high-pressure steam engine, which in the long run led to the development of the great big pumping engines that Cornwall became famous for.'

The main market for the steam engine was industry and other engineers took up Trevithick's application of high-pressure steam. Cornish engines became famous the world over and during the course of the nineteenth century they got bigger and bigger. The one you can see at Cornish Engines at Pool is the last of the line of Cornish pumping engines. Fred said:

It was erected in 1924 and it represented the ultimate in mine-pumping engineering that had all started way back in the days of Newcomen. It ran on a three-shift system with a team of three engine drivers day and night. It burned 50 tons of coal a week; it has a 90-inch diameter cylinder and a 10-foot stroke. It's incredible. The majority of these great engines were actually made in Cornish foundries by people like Holman Brothers and Harvey's of Hayle who made this very engine on the Taylor shaft at Pool. By the early days they were exporting them all over the world; then the Cornish engineers went out and erected the things and quite often stayed out there to work the mines as well.

With the development of these engines the landscape not just of Cornwall, but of the rest

Another great idea that Richard Trevithick came up with was the chimney

of Britain, was changed dramatically. It was another of Trevithick's inventions, as Fred describes, that had the most dramatic effect:

> Another great idea that Richard Trevithick came up with was the chimney, which improved the draught on boilers and eventually became quite common in all industrial areas on the skyline; not least in Lancashire where I come from. One of the very first sights that I remember when I were a small boy were all the factory chimneys that surrounded Bolton, especially all those near Bolton Wanderers' old football ground at Burnden Park, where I lived. When the sun was setting in the west, I used to look at them silhouetted against the evening sky above the area that I played in as a kid. The chimneys towered above everything and you've got to remember that at the bottom of every one of them there was a great big boiler that raised the steam that provided all the power for the mill.

One of the reasons that there were so many chimneys around Bolton was the proximity of the town to the Lancashire coalfields. In the late eighteenth century and the first half of the nineteenth the presence of large coal deposits in parts of England like Lancashire was one of the most important contributing factors to the development of steam and the Industrial Revolution. The unlocking of British coal seams turned Britain into a great manufacturing state. The steam engine provided the power to drain the coal mines, which supplied the fuel that was needed for newly developed methods of smelting iron; which then, in turn, provided the metal used in the construction of engines and machines. Until the advent of the steam engine most manufacturing work was done in small cottages and workshops. It was in England that the first big industrial towns, crowded with tall smoking chimneys and noisy factories, were built. The earliest factories were built for only ten or twenty people to work in, but by the middle of the nineteenth century, thanks to the power of steam, a single factory might employ hundreds of workers and operate many different types of machinery.

In factories the steam engine gradually took over the role of the waterwheel. The power from a single steam engine or a series of engines was used to power dozens and dozens of machines in places like textile mills and engineering works. Power was transmitted by hundreds of yards of line shafts and belts. The power of the steam engine passed through horizontal shafts and toothed gearing to vertical shafts. These, through yet more toothed gearing, drove horizontal shafts on each floor of the factory or works from which individual machines were powered by leather belts and pulleys.

The Italian-inspired, campanile-style, square chimney of India Mill dominates Darwen

India Mill

As steam engines got bigger, mills and factories got grander in scale. One that always impressed Fred was the India Mill at Darwen between Bolton and Blackburn. Most of the building was completed in the mid-1860s and in 1871 two 125 hp W. & J. Yates beam engines were installed in the engine house next to it to drive the machinery. No expense was spared on the mill and today it is still a very impressive Grade II listed building constructed of ashlar and par-point masonry throughout, with ornate decoration to cornices, door surrounds and window openings. But the mill's crowning glory is its chimney. The Italian-inspired, campanile-style, square chimney dominates the whole of Darwen and can be seen from miles around. The massive base has recessed round-headed panels set in rusticated stonework and a stepped stone plinth. Above, the brick shaft is detailed with inset blue and yellow panels, slit windows and blind arches. The windows led many people to believe that there was a staircase inside the chimney, but this is not the case; the windows and arches were there purely for decoration. The top terminates in two prominent overhanging cornices, the lower of which has stone urns at the corners. When it was completed in 1867, the chimney was the tallest and most expensive in the whole of Britain.

India Mill reaches 300 feet in height

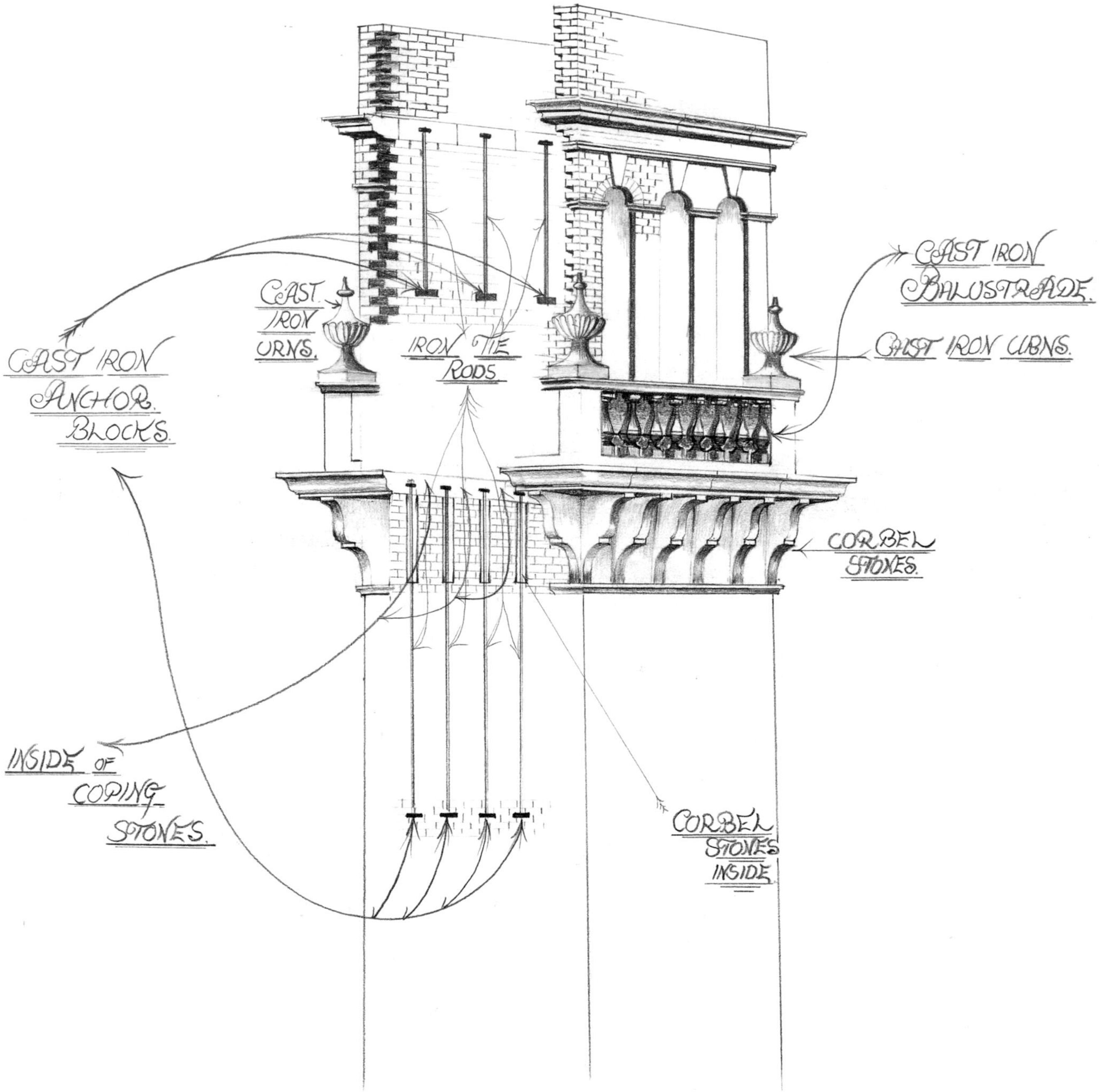

The man who designed it must have had a vision: he must have seen the tower in St Mark's Square in Venice and based his designs on that

The mill – and particularly the chimney – was a wonderful example of the flamboyant confidence of the Lancashire cotton industry in the middle of the nineteenth century. The story goes that the ownership of the land where the mill was built was disputed between the Shorrock and Hilton families. It was agreed that the land would go to whomever came up with the most impressive plan for building on it. Cotton magnate Eccles Shorrock won hands down over his rivals, who had a paper-making business. The opening of India Mill in May 1868 by the Marquis of Hartington was a very grand affair with lords and ladies and a vast exhibition of paintings and sculptures covering three floors. Many of the paintings, by famous artists like Gainsborough, Van Dyck and Dürer, are today housed in some of the world's leading art galleries and are worth millions. For the lads who actually built the chimney with their handmade bricks there was a more modest lunch at the nearby Crown Inn. Little could they have imagined at the time that one day their work would occupy such an important place in the annals of industrial architecture. We went to look at it with Fred:

> This must be one of the grandest chimneys in the whole of England. It shows that when the Victorians designed things they didn't just have to be functional; they had to look good as well. The man who designed it must have had a vision; he must have seen the tower in St Mark's Square in Venice and based his designs on that. I've seen pictures of the one in Venice and this is just like it. It's all very ornate, and when they first built it, there was more ornamentation than there is now. There was a whole lot of ironwork that was right up on the top of it, but this was taken down in the 1930s by a local steeplejack, who became a good friend of mine, to provide ironwork for the war effort. All that ironwork on the top stood on 10 feet of brickwork which was built on the upper projection that overhangs by about three feet. Below this top overhang there's about 35 feet of very ornate chimney stack, then there's a ledge with 4 finials, one on each corner. Each of these is 8 feet high, made from stone topped with cast iron. Then below these you get the biggest projection which overhangs by about 5 feet and slopes down about 18 inches. At this level there are 32 cast-iron balustrades about 3 feet high built into the brickwork. Behind these balustrades some peregrine falcons have made their nests. They eat all the local pigeons and they make life difficult for anybody who has got to go up there to do any conservation work.

Some years ago I had the pleasure of laddering this chimney for an inspection

Over the past thirty years or so many of Lancashire's finest chimneys have been reduced to rubble as cotton and spinning mills have closed. India Mill itself closed in 1991 and has now been preserved as a business centre. Soon after its closure, Fred had a look at the structure.

Some years ago I had the pleasure of laddering this chimney for an inspection. It looks pretty impressive when you are down at the bottom of it but when you get up there, some of the stones up at the top are massive; they must weigh maybe 5 or 6 tons each. When they built it, it is reputed that they had a roast beef dinner at the top to celebrate. I don't know whether they did or not, but it's certainly big enough up there. You could ride a bike round the top.

They're funny things chimneys; some feel friendly; others feel nasty. This one feels friendly. In spite of those big overhangs, I wasn't scared of it at all. I've been on much lesser chimneys than this where I've felt more scared. But if you do put a hand or a foot wrong it's half a day out with the undertaker. It's a long climb to the top and the last time I did it I was turned fifty, so it's a bit knackering. On the way up you can see at close quarters all the bands of brickwork. They must have looked nice when they first built it – all those yellow and blue and red bricks. The biggest overhang is the first one you come to and it sticks out 5 feet from the face of the brickwork. It means you've got to go out backwards to get round the corner, more or less hanging out in mid-air. Overcoming that sorts out the men from the boys. Harry Holden, the famous Lancashire steeplejack, used to say to his lads you need dack on your shoes to get round this corner. That's glue, for those who don't know. When you get up and over on to that cornice it all looks very impressive. The plant pots on the corner are made of cast iron. It's

They're funny things chimneys; some feel friendly; others feel nasty. This one feels friendly. In spite of those big overhangs, I wasn't scared of it at all.

unbelievable the pains they went to so they could make it look good.

The modus operandi for building it would have been all from the inside and all built leaning over. It would all have been built from a platform in the middle and as the walls went up the platform would be moved at 6-foot centres up the middle of the chimney. But when they got to the top they no doubt had a steam winch to pull the great big stones up. The overhang on top were held in place by great vertical tie rods down the middle anchored into the brickwork below so they couldn't fall off. They're still there, them iron rods inside the chimney. When you look down the middle you can see this mass of iron work that holds all the big overhanging bits on. I wouldn't like to have to dismantle it the same way as they put it up; pretty difficult, I would say. The top of the chimney in my drawing is a bit exploded in a way, so that you can see the tie rods in the middle. When I first got on top and looked down the middle I was fascinated. The great corbel stones that you can see are 5 or 6 feet tall and about a foot thick. When they got them up here, they'd have them precariously balanced on wooden scaffolding. Then of course they'd put internal slabs of iron in holes chiselled into the inside of the chimney and on top of them there'd be little nuts to fix the stone on to. Further down the inside of the chimney there are cast-iron bricks built into the brickwork proper – like an anchor to stop the stones tipping off the edge.

The last time Fred laddered the chimney was so that somebody could abseil down it to raise money for charity – not something that Fred regarded as a proper steeplejacking job. He found it all a bit strange. 'The guy insisted on having a rope tied round him as he was climbing to the top because he said he didn't feel very safe on the ladders. But once he'd got off them and he was dangling in the air on the end of his rope, he was happy!'

In the Lancashire mill towns little is left standing now of their proud industrial heritage. Surrounding chimneys, much less grand, which once dotted Darwen by the dozen steadily disappeared in the sixties and seventies and India Mill chimney now stands tall in proud isolation as a lone memorial to King Cotton. 'The Industrial Revolution,' Fred said, 'was a major chapter in our history and industrial heritage sites like this that have been preserved are as important as our great houses, castles, abbeys and cathedrals. It's all so sad, though, that places like this were still working in our lives, but alas they've nearly all gone now.'

The Steam Locomotive

Riding on a replica of Trevithick's Puffing Devil

Fred is filmed driving Tom Broaden's replica of Trevithick's road carriage

The development of the steam engine and the impetus that it gave to industry brought rapid growth to the new industrial areas of Britain, as Fred described:

> In Cornwall the advances Richard Trevithick made in pumping engines and winding machinery gave the county an unbelievable prosperity between 1800 and 1870. Today, though, the whole industrial landscape in Cornwall is a bit sad. Just about all derelict now – with very little trace of the work of Trevithick, who was one of the greatest pioneers of steam. Trevithick's engines consumed three times as much coal as Watt's but they were compact, simple and easy to install. They also didn't need copious amounts of cold water which was needed by condensing engines. The advances Trevithick made in steam power for pumping the tin mines and for winding ore from them certainly helped make the period 1800 to 1870 a time of unbelievable prosperity for Cornwall. But his development of the Cornish engine wasn't the only thing that made him one of the giants of steam. It is as the first engineer to apply steam power to the haulage of loads on a railway that Trevithick made his greatest contribution to technology. It's my opinion that Trevithick never really got true recognition for his

Richard Trevithick

Fred had a look at a replica of Pen y Darren *at Ironbridge Gorge Museum*

contribution to the development of the steam engine – not only in mining engines but in the field of steam-powered road transport and railways as well – because it was his invention of high pressure or strong steam as he called it that led to the development of some of the world's first steam-powered locomotives. His first was designed to run along the road.

Trevithick's Road Locomotives

James Watt had been applying his inventive mind to the problem for many years. He wanted to apply his engine to locomotion, using either a non-condensing engine or an air-surface condenser. He included the locomotive engine in his patent of 1784 and his assistant Murdoch made a working locomotive. It is said to have run at 6 to 8 miles an hour. After seeing a model engine constructed by Murdoch, Trevithick, who was his pupil, was determined to build a carriage to run on common roads. When Watt's patent ran out in 1801, Trevithick built a prototype road locomotive at Camborne in Cornwall. On Christmas Eve 1801 he ran this up a hill for several hundred yards with a number of people hanging on to it. This road locomotive was named the *Puffing Devil*, but unfortunately it burnt out while Trevithick and his friends were celebrating their success at a nearby inn.

The following year, Trevithick took out a patent for a passenger-carrying steam road carriage. It was assembled at Felton's carriage works at Leather Lane, London. The engine may have been tested in another machine called the Tuckingmill locomotive which was reported to have been stuck on the road between Camborne and Redruth 'because its wheels could not get sufficient grip of the road'. The steam carriage resembled a stagecoach, and had four wheels. It had one horizontal cylinder, which, together with the boiler and furnace-box, was placed in the rear of the hind axle. The motion of the piston was transmitted to a separate crank-axle, from which the axle of the driving wheel derived its motion. The steam-racks and force-pumps and the bellows used in generating combustion were worked off the same crank-axle. It was the first successful high-pressure engine constructed on the principle of a piston moving, by the elasticity of steam, against the pressure of the atmosphere, and without a vacuum. The piston was not only raised but also depressed by the steam.

On completion, the London Steam Carriage was driven about ten miles through the streets of London to Paddington and back through Islington with seven or eight guest passengers. This was the first trip of a self-powered passenger-carrying vehicle in

the world. However, disaster struck again when Trevithick and his colleague crashed the carriage into some house railings. Ultimately, the project to build a steam-powered passenger-carrying vehicle was not a success. It was too expensive and it needed two men and a bag of coal to do what one man and a bag of hay could do with a horse-drawn vehicle.

Rail Locomotives

But Trevithick didn't give up on the idea of applying steam power to locomotion and he now turned his attention to developing a steam carriage or locomotive to run upon the tram-roads that were by now in general use in Britain. It was in 1804 that he started building his machine called the *Pen y Darren*. The design was based on a compact stationary engine/boiler that he had developed and it was completed and tested within the same year. The boiler was cylindrical in form, flat at the ends and made of cast iron. The furnace and flues were inside the boiler, in which a single cylinder of 8 inches in diameter and 4 feet 6 inch stroke was immersed upright. It hauled a load of 10 tons of iron and 70 men along 10 miles of tramway at the rate of 5 mph at the Merthyr-Tydfil Railway in South Wales, winning him a five-hundred-pound bet for building the first locomotive in the world.

But, despite this initial success, the railway was abandoned as a failure. The first attempt to adapt the steam engine to work on a railroad had worked but *Pen y Darren* was slow and heavy and kept breaking the tracks. The track on which Trevithick ran it had been a horse-drawn tramway to transport iron to the Glamorganshire Canal. The rails were made of cast iron so they

On the Pockerley Waggonway at Beamish Open Air Museum, where they've recreated what the railways of the early 1800s looked like

were brittle and they cracked under the weight. The experiments stopped and the tramway returned to horse-drawn trains with smaller loads. Advances in rail technology were needed: new materials of wrought iron and later steel; and different shapes – rolled iron with the flange taken off the rails and put on the train wheels. There were problems, too, getting sufficient grip for smooth wheels to run along a smooth track.

Trevithick built a similar locomotive at Gateshead in 1805 and in 1808 he demonstrated a third on a circular track laid near Euston Road in London. On 21 February 1808, one of the most significant dates in railway history, he put to work a steam carriage called *Catch-me-who-can* on which he gave novelty rides. It was a very plain and simple machine. The steam-cylinder was set vertically in the after-end of the boiler, and the cross-head was connected to two rods, one on either side, driving the back pair of wheels. The exhaust steam entered the chimney, aiding the draught. This engine, weighing about 10 tons, travelled at 12 to 15 mph on the circular track. Eventually the engine was thrown from the track because of a break in the rail and, because all of Trevithick's funds had been spent, it was never replaced. He then abandoned these projects because yet again the cast-iron rails proved too brittle for the weight of his engines.

Other engineers, however, worked away at the idea and in the early 1800s one place led the world: Northumbria. On the Pockerley Waggonway in the Beamish Open Air Museum they've re-created what the railways of the period actually looked like. Inside an engine shed, which is a copy of Timothy Hackworth's shed at Shildon on the Stockton and Darlington line, there's a collection of locomotives from the very earliest days of the railways. The magnificent old loco that Fred travelled on when he was there is a full-size replica of *Steam Elephant*. 'It's so old that it's got a wooden chassis,' said Fred. 'And when you study it and you look at the funnel it's obviously where it got its name from. It's just like an elephant's trunk. It was originally built in 1815 by Chapman and

The magnificent old loco that Fred travelled on at Beamish is a full-size replica of Steam Elephant

Buddle for the Wallsend Colliery and it worked from 1815 to 1840. Then it mysteriously disappeared.'

These early locomotives like *Steam Elephant* were all built for the coal mines in the north-east. It was in these mines that the most famous man in railway history worked. When he was a young man George Stephenson was the enginewright at Killingworth pit so he would have been very familiar with locos like *Steam Elephant*.

George Stephenson

Stephenson is often referred to as the inventor of the locomotive. This isn't strictly true but he did play the leading part in turning the invention into a practical means of hauling coal and transporting passengers over long distances. It was the beginnings of the railways as we know them. Stephenson built his first engine, *Blücher*, in 1814 while he was working at Killingworth, but it proved to be defective and the cost of using

It must've been quite exciting really, like being an airline pilot in 1825

The original Locomotion at Head of Steam, Darlington Railway Museum

it was about as great as that of using horse power. Stephenson was determined to build another engine, this time to a different plan, and he patented his new design in 1815. It proved to be a much more efficient engine than *Blücher*. By this time he was being widely acclaimed as a brilliant and ambitious engineer. His strong advantage was that he combined a great natural inventive talent with an excellent mechanical training and an entrepreneurial spirit.

In 1823 Stephenson was made engineer for the Stockton and Darlington Railway, which was planned to transport coal from the Durham coalfields to the sea. Edward Pease, a Quaker industrialist from Darlington, was one of his main backers and he was a great advocate of steam power. He could see that steam would bring immense advantages and he was so keen on getting steam locomotives to provide the power on the new line that he gave Stephenson an advance of a thousand pounds to help him to begin the business of locomotive-building at Newcastle. The building of Stephenson's Forth Street Works, which went on to become one of the most famous establishments in the history of railways, was begun in 1824.

On 27 September 1825, a train with a strange-looking engine in front and thirty-four wagons filled with passengers, flour and coal behind, steamed into history. For many people this is the beginning of railway history. George Stephenson's *Locomotion No. One* pulled the train from Shildon to Darlington and then on to Stockton. Fred described the momentous journey:

> The whole train weighed 90 tons and went at the unbelievable speed of 12 miles an hour. It had two cylinders that drove cross-beams and connecting rods. The locomotive ran on four wheels, with a four-wheel tender to carry the coal and water. The driver had to balance precariously on a platform on the left-hand side of the engine, where he could control the supply of steam to the cylinders and operate the primitive valve gear. The fireman rode on the front of the tender, although when he was not stoking the fire he could ride on the platform on the opposite side of the engine to the driver. The engine, like many of the early locomotives, was not fitted with any brakes. On the first run George Stephenson actually drove the locomotive and his brother acted as fireman. It must've been quite exciting really, like being an airline pilot in 1825. Incredible! It had no brakes to stop the thing and the fireman had to actually jump off and pin down the brakes on the coal wagon. Quite a hairy occupation! On the way to Darlington the train frequently reached a speed of 12 mph, and on one occasion it got up to

15 mph. People on horseback and on foot tried to race the train as it passed triumphantly along. Spectators came out in their thousands to line the track, waving and cheering as this strange contraption went past. When they got to Darlington it seemed the whole town had turned out to see the train arrive, and then steam off towards Stockton, to arrive there in just over three hours.

After the events of the opening day the railway had to settle down to work to earn money and to do what it had been planned to do. A second locomotive was delivered on 1 November 1825. This had been built by Stephenson's son's company Robert Stephenson & Co., and two more followed in 1826. The first two locomotives cost five hundred pounds each and the second two cost six hundred each. Although, inevitably, there was some trouble with these early engines, the steam engine soon proved its superiority over horse-drawn transport, particularly because of the much greater loads it could handle. As a result the company persevered with steam locomotives and, largely thanks to Timothy Hackworth, their faith in steam was eventually justified.

Hackworth had been appointed Locomotive Foreman on the Stockton and

Darlington Railway in May 1825 at the age of thirty-eight. He started on a salary of £150 a year plus a free house and coal and it was on his shoulders that the task of keeping the engines in running order fell. To do this he only had very primitive facilities and just a few staff. He established workshops at the terminus of the line in Shildon and gradually built them up to such a high standard that new engines could be constructed and major repairs undertaken there. In 1833 the Stockton and Darlington Railway Company decided that the working of the locomotives should be looked after by contractors and Hackworth, amongst others, agreed to take on the work. At the same time he was able to establish his own locomotive-building works at Shildon.

When they set up their railway in the 1820s the founding fathers of the Stockton and Darlington Railway couldn't have foreseen that within fifteen years of its opening the whole country would be feverishly building lines, or that within forty years rail transport would have become the principal means of moving passengers and goods. The one that really set the ball rolling was the Liverpool and Manchester Railway. A railway between these two rapidly developing northern cities had been projected at about the time that work on the Stockton and Darlington had started and the Liverpool and Manchester Bill had been carried through Parliament. Stephenson was involved from the outset and, not surprisingly, he urged the use of locomotives rather than horses. The line was built with Stephenson as principal constructing engineer. As it approached completion it became necessary to settle the long-put-off question of what sort of motive power was going to be used for the line.

Some of the directors and their advisers were still keen on the use of horses; some thought stationary hauling-engines were preferable; and the remainder were undecided. Most of them, however, were worried that the only locomotive available to them was one called *Agenoria* which was too slow and inefficient. Stephenson was almost alone in backing the locomotive, but after long debates he persuaded the board to give the travelling engine a chance and a competition was set up with a prize of five hundred pounds for the best locomotive engine. This reward was printed, published and circulated throughout the kingdom, and a considerable number of engines were constructed to compete at the trial. Only four engines, however, were finally entered for the trial which took place on a stretch of the new line at Rainhill. The entries included an engine built by Timothy Hackworth called *Sans Pareil* and George Stephenson's entry, *Rocket*.

Stephenson's revolutionary new design, which incorporated the fire-tube boiler, can be seen in this cutaway replica at the National Railway Museum

Rocket

Stephenson had come up with a revolutionary new design that increased the power and speed of the engine relative to its weight. It was a major breakthrough and marked one of the key advances in railway technology. It also confirmed Stephenson as one of the premier engineers of his age and as a major engineering contractor for the emerging railway network. *Rocket* ran at speeds of between 25 and 30 miles an hour, drawing a single carriage carrying thirty passengers. This success settled completely the whole question of what motive power should be used and the Liverpool and Manchester Railway was at once equipped with locomotive engines.

With the help of his drawing of *Rocket* Fred explained the significance of the design:

> Stephenson went way off track and came up with a brand-new revolutionary design which of course incorporated the fire-tube boiler and which really is the prototype for all modern locomotive boilers that we know today. The thing is, in relation to its weight and the power it had, it went much faster than any other locomotive that had been built before. It did away with all the beams and the levers of the earlier locomotives. This at the time was a revolutionary boiler. The fact that it had twenty-five copper tubes going from one end to the other is something that

Fred's drawing of Stephenson's Rocket

had never been done before and the way the heat was transferred from the fire into the tubes were the beginnings of the true firebox. It would have worked much better than just a single fire tube into the boiler which is all there had been on earlier locomotives. The other wonderful thing were the blast pipe. When the piston had gone up and down and turned the wheels round, the escaping steam from the valve chest went along the copper pipe we can see and into the base of the funnel where it created a vacuum in the bottom, which drew the fire with an unbelievable degree of violence. The other wonderful thing were the connecting rod which connected the piston directly to the crank pin on the hub of the front wheels which led to nice smooth running.

The new features that Stephenson had put into his engine were to become standard for all steam locomotives. In the boiler hot gases from the burning coal were drawn through the twenty-five tubes which were surrounded by water. This helped to boil the water more quickly which in turn made faster running possible. His biggest innovation, the direct drive from the cylinders and pistons to the wheels, eliminated the need for the complicated gears and levers of every other locomotive designed up to this point. Knowing that it would improve the air draught through the firebox Stephenson also introduced the blast pipe, which sent exhaust steam up the chimney. The powerful draught creates the familiar 'chuff chuff' of the steam locomotive. Until Stephenson came along with *Rocket* everything else, including its main rival *Sans Pareil,* had been a beam engine on wheels.

Stephenson went way off track and came up with a brand new revolutionary design

Sans Pareil wasn't a patch on *Rocket*; the driver and fireman were at opposite ends of the locomotive and heat from the fire passed through the length of the boiler in one direction and then all the way back again, using old technology which wasn't as efficient as Stephenson's multi-tube boiler. *Sans Pareil*, however, was also a victim of bad luck. Timothy Hackworth managed to build his locomotive even though he didn't have a workshop. He had to get all the parts manufactured outside by independent contractors (the main parts, the cylinders, were actually done by his rival George Stephenson). Then when it came to the competition the boiler almost boiled dry when the water pump failed and one of the cylinders burst. Some people believe that Stephenson sabotaged his rival.

Rocket was the first steam locomotive as we know them – the 'granddaddy' of them all. It paved the way for every other steam

Timothy Hackworth's Sans Pareil *was one of the* Rocket's *rivals at the Rainhill Trials*

locomotive constructed right up to the 1960s. For some years after this, his first great triumph, George Stephenson gave his whole time to the building of railways and improvements to the design of his engine. He was assisted by his son, Robert, who eventually took over the business after his death. By 1830 around a hundred locomotives had been built in Britain and Robert Stephenson introduced the *Planet* class for work on the Liverpool and Manchester Railway.

George Stephenson is known as 'the Father of the Railways' but his early locomotives were quite primitive, spindly-looking things. The man who transformed them into big, powerful engines capable of transporting hundreds of people at high speeds was Robert. In the early years of Queen Victoria's reign Robert Stephenson was in the forefront of creating a railway network which was to transform the lives of millions. By the 1840s the small workshops that had been set up by men like the Stephensons were getting bigger and more industrialized. Their Forth Street Locomotive Works at Newcastle where *Rocket* had been built was expanding rapidly and Robert Stephenson soon turned it into the biggest locomotive manufacturer in the world. Customers from far and wide would come here to discuss their requirements with Mr Stephenson. Rapid advances were made in a very short time and not just in locomotive-building. Robert Stephenson and his company also built the lines and the bridges and all the engineering works involved in the construction of a full-size railway. As well as building the machines that changed the world the great Victorian engineers like Robert Stephenson left us with engineering wonders that changed the landscape for ever.

The original Sans Pareil *and the replica built for the museum can be seen at Shildon Locomotion Museum*

CHAPTER FIVE

ENGINEERING WONDERS

As Fred Dibnah was growing up in Bolton he was surrounded by canals and railway lines, bridges and tunnels. He was always fascinated by great civil-engineering projects like this, and by the lives of the men who changed the landscape of Britain for ever.

Just near the house I lived in when I was a small boy in Bolton there was a rather splendid Victorian railway viaduct. It was round the back of Burnden Park where Bolton Wanderers used to play. It consisted of about five stone pillars and, as a kid, they seemed to me to be made of the biggest stones in the world. Each one was about 5 feet long and 2 feet deep and 3 feet wide; they were massive.

Between each of the stone pillars there were iron girders that the railway track ran on and you used to be able to get on this railway viaduct at one end and walk right across over a big valley, which seemed to be about a million miles below when you were only six years old. Then when I got to about eight years old I used to go climbing in the iron girders underneath the track and when a locomotive came along with a load of coal wagons on, the whole lot used to shake about. They were really quite exciting times in a way. It was magic, and those memories have never left me.

Bridges

From the very earliest times man has had the problem of getting across rivers. Over time, bridges became less basic but designers were always limited by the length of the arch. All that was to change in the eighteenth century when the world's first iron bridge was built across the river Severn at Ironbridge in Shropshire.

Advances in the manufacture of wrought iron in the eighteenth century made it possible to build iron bridges

Ironbridge

When tenders were invited for the construction of the bridge in 1776 only one man was willing to take on the challenge: Abraham Darby III. The problem was that an exceptionally long central span – 120 feet – was needed to reach across the gap. Other engineers feared that stone wouldn't be able to do the job and they had no alternative solutions. But Darby did. For more than half a century his family and their Coalbrookdale Ironworks had been in the vanguard of technological change. The first Abraham Darby paved the way for the Industrial Revolution when, in 1709, he successfully smelted iron for the first time with coke and made Coalbrookdale into the cradle of the industrial age. They made the first cast-iron cylinders for steam engines in 1722; the first pig iron for forge work in 1748 and the first cast-iron rails for the horse-drawn predecessors of steam railways.

Ironbridge Gorge, Shropshire – the world's first iron bridge

BRIDGE WAS CAST
COALBROOK DALE
AND ERECTED IN THE YEAR

Thomas Telford built his suspension bridge over the Menai Straits between 1819 and 1825

It was the success of the Darby family that led to the great expansion of the iron trade in the eighteenth century. Later ironmasters in the Gorge used iron to make the first iron railway wheels, the first iron rails and the first steam railway locomotive and visitors came from all over the world to learn about the new technologies.

It was Abraham Darby III who planned the construction of what was the world's first great iron bridge for which the area was renamed. In the 1750s, there were about six ferry crossings operating within the Gorge, carrying raw materials for the industry and the workforce. But as industry's requirements grew, so did the need for a better way of crossing the River Severn.

In 1773 Thomas Farnolls Pritchard, a Shrewsbury architect, wrote to John Wilkinson, a local ironmaster, suggesting a new bridge made entirely of cast iron should be built across the river. As the leading businessman in the Gorge, Darby was brought in as the financial backer of the new bridge.

But there were doubts about whether an 'iron bridge' would work. Although it was known that small iron bridges already existed in China, there were none of such a mass and size as the one that was planned to span the gorge.

Abraham Darby III's workmen raised the huge arches of the iron bridge in the summer of 1779. Each of the ribs that made up the arches weighed 6 tons. Neither bolts nor rivets were used, as parts were slotted together and fixed with wedges.

The town of Ironbridge became a prosperous industrial community in the late eighteenth century and by the nineteenth century it had been visited by many famous people. The bridge itself certainly paid its way. Traffic continued to use it until 1934. Today this elegant structure still remains, the first important example in the world of the structural use of cast iron.

Thomas Telford

The great civil engineer Thomas Telford built several arched cast-iron bridges. Born in Westerkirk in Scotland in 1757, Telford was a qualified stonemason and self-taught engineer who became a great builder of suspension bridges. Suspension bridges were new. The basic principle of suspending a path or a roadway from cables rather than building one on top of arches meant that wider spaces could be crossed. The idea was taken up very rapidly, but it was not until the 1820s that advances in the design and manufacture of wrought-iron chains made it possible for Telford to build his two great suspension bridges.

Bridges were needed at Conwy and over the Menai Straits to get traffic to Holyhead

for the Irish ferries. It was this that led to the setting up of the Holyhead Road Commissioners in 1815 and, over the next fifteen years, with Thomas Telford as their principal engineer, they built a road from Shrewsbury to Holyhead which was regarded as a model of the road-making art. Although it runs through the heart of Snowdonia, Telford insisted that it should never exceed a gradient of 1 in 20. This permitted a stagecoach to maintain a steady 10 miles per hour. Telford's road crossed the river Conwy at Betws-y-Coed by the Waterloo Bridge. It has a single cast-iron span of 105 feet with pierced spandrels elaborately decorated with national emblems. The bridge still carries modern traffic. Further down the road Telford was faced with a more formidable obstacle: the Menai Straits. Because of the width of the straits Telford proposed that he should build a suspension bridge and in 1818 money was granted for the structure which was to become his most famous and important work.

The Menai Straits Bridge

Work started on the bridge in 1819 and by 1824 the stone arches and towers at either end of the bridge had been completed and they were ready to put the chains in place that would support the roadway. A 400-foot-long raft was built to carry the first chain from its mooring to the main towers and, once the raft was in place, blocks were fitted to the chain and thirty-two men lifted it up and over the Anglesey tower, allowing the raft to float away. Over the next ten weeks all fifteen chains were put in place and on 9 July 1825 a huge ceremony took place to celebrate the achievement of linking the island of Anglesey to the Welsh mainland. When it was opened in 1826 the Menai Bridge was by far the longest suspension bridge in the world.

The Menai road bridge was originally of cast iron, supported by sixteen wrought-iron chains passing over tall masonry towers on either bank and anchored below ground. Telford's original chains have now been

The wrought iron chains on Conwy Suspension Bridge are held together by fish plates spaced in between them

replaced by steel ones, so Fred went to have a look at the smaller Conwy Bridge where all of the original wrought iron is still in place.

The Conwy Bridge

The historic suspension bridge next to the castle at Conwy was designed by Telford as part of the great highway between Chester and Holyhead. It was built at the same time as the one over the Menai Straits and both were opened for traffic in 1826. The Conwy Bridge is smaller and because of this and its more sheltered position it wasn't exposed to the same hazards.

For a bridge-builder, the most important consideration is always the foundations. Telford selected the crossing under the castle because the solid rock on which the castle stood would provide good foundations for the bridge abutments. About a hundred yards across the main river channel from the castle lay a rocky island, which would provide the base for one of the towers of the suspension bridge. This island would be reached from the eastern shore by a 660-yard embankment, the construction of which Telford later described as being one of the most arduous operations in North Wales. Work started from the east bank of the river and for three years labourers moved thousands of tons of hard local clay to build the embankment. The first stone for the bridge itself was laid in April 1822 and by June the suspension towers were already raised several feet. The stonework in the towers is castellated to harmonize with the

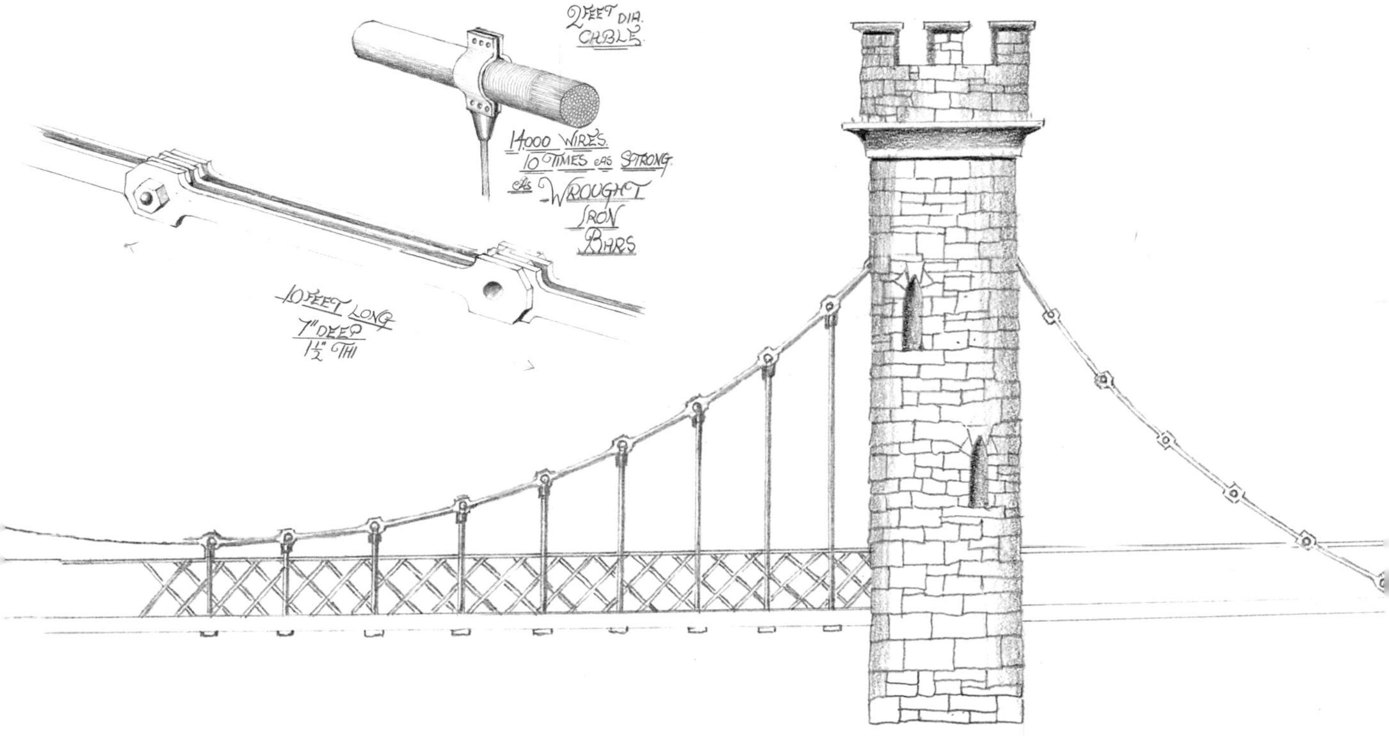

architecture of the castle. The chain supports consist of pairs of solid stone turrets linked by a wall containing an arched opening 10 feet wide. By February 1825 the masonry was completed as were the chamber anchorages for the suspension chains.

The trickiest part of the building work was the fitting of the wrought-iron chains which had been made at workshops in Shrewsbury. Fred's drawing shows how each chain consists of 5 bars about 10 feet long and about 4 by 1¼ inches thick, with an eye forged at each end. 'They're all held together,' said Fred, 'by fish plates that are spaced in between them and then two great bolts slammed through the lot about 3 inches diameter.' The chains are arranged in two tiers of five links joined by deeper plates. The joints are alternated and from the junction plates vertical rods at five-foot intervals carry the roadway. To get the chains into place the method used at Menai of hoisting them up in one piece from barges had to be abandoned because the current was too strong. 'So,' Fred explained, 'they got the chains across in rather an unconventional way. Link upon link was joined together up on a rope bridge stretched between the towers. Once they'd got the chains across, it were quite a simple job putting the vertical bolts and bars down to the road surface and building the road on it. It took little more than four years to construct and by 1 July 1826 Conwy Bridge was opened. Along with the Menai Bridge, it was a really great engineering achievement in its day.'

Canals

Fred always had a great interest in canals and the way they were built

Telford's Pontcysyllte Aqueduct carries the Shropshire Union Canal over the River Dee near Llangollen

Pontcysyllte Aqueduct

Thomas Telford was one of our greatest civil engineers. He built roads, bridges and canals, but one of his most dramatic engineering feats is the 1,000-foot long Pontcysyllte Aqueduct, which carries the Shropshire Union Canal across a valley high above the River Dee near Llangollen in North Wales. Work on this aqueduct, which spans the Dee Valley on eighteen piers of exceptionally fine construction, began in July 1795 and it took ten years to complete. It is built of local sandstone, expertly cut and dressed by Telford's masons and bound together by mortar joints of unusual thinness. The aqueduct rises to a total height of 116 feet and there are 19 arches each with a span of 45 feet. The waterway itself is carried over in a cast-iron trough 11 feet 10 inches wide and 5 feet 3 inches deep. This is constructed of cast-iron flanged plates bolted together – each casting dovetailing into the next with the joints sealed with lead and Welsh flannel.

In order to test the trough after construction it was filled with water from a nearby brook and left to stand for a period of six months. During this time it was found that it did not leak a drop. The opening ceremony took place in November 1805 in the presence

I have always had a long, interesting relationship with this bit of canal, believe me

of eight thousand people and six boats went over the aqueduct from north to south and back again. The canal is still open today and you can take barge trips over the aqueduct. 'It's quite spectacular,' was Fred's verdict, 'and, even for somebody like me with a head for heights, it can be a bit unnerving, because it feels as though the barge you are on is going to float right off the edge.'

The Manchester-Bolton-Bury Canal

Fred always had a great interest in canals. As a small boy he used to spend many an hour along the banks of the Manchester-Bolton- Bury Canal. Later in life he became the honorary president of the canal society and a firm supporter of a scheme to reopen the canal. It was built towards the end of the eighteenth century to link the towns of Bolton and Bury with their larger neighbour Manchester. During its working life it carried mainly coal mined from the many collieries in the area – particularly Ladyshore Colliery which stood on the banks of the canal. Fred reminisced:

> I know this canal very well. All my life I've played around it and I've even sailed down it in a homemade boat. I've ridden my bicycle along the edge of here all the way from Bolton to Bury and I can't even swim. I have had a long and and interesting relationship with this bit of canal, believe me. One of the earliest memories that I have of my interest in industry is as an eight-year-old riding along the towpath on the canal on my bike with my dad, when half my concentration was on making sure I didn't ride off the towpath. All along the banks of this canal at that time there were wonderful examples of Victorian mining engineering, all made of wood, with inclined railways down the side of the banking, pit-winding headgear, a huge boilerhouse at the bottom of a great big chimney and near the access to the pithead a little basin for the canal boats. It was like going into a time warp.
>
> The canal was part of a whole network around Manchester and Bolton that had been built in the eighteenth and nineteenth centuries to transport coal. But it wasn't just coal. Every conceivable type of cargo was transported along here – raw cotton and finished cotton goods; timber, bricks and stone for the building industry; food; manure for the farms – because even I remember when there were all fields with cows round here. And there was a passenger service as well. It was a nice comfortable way to travel, but there was the occasional disaster. Once a party of drunken passengers going from Bury to Bolton started brawling and fighting amongst themselves. It was so bad they finished up

The Bridgewater Canal at Worsley, Lancashire

capsizing the boat and several people including two children were drowned.

Sadly the canal comes to an end now before it reaches Bolton so it can't be used but within a stretch of a couple of miles not far from Bolton you can see the massive engineering achievements of the eighteenth century when they were doing this sort of thing for the first time, and because it is now disused and semi-derelict it's easier to see how it was built, especially in sections where the water has been drained from it. You can walk along the bed of the canal and see the quality of the stonework, even below water level, and they didn't lessen the quality of the workmanship even as they got to the bottom.

The canals were like the arteries of the Industrial Revolution. They helped to provide cheaper goods and raw materials and cut travelling time down massively. Without them the Industrial Revolution couldn't have happened and it all started off just down the road from me in Worsley in Lancashire.

In the second half of the eighteenth century, Britain was bursting with industry and commerce and a way had to be found to move raw materials to the new factories that were springing up and to get their products to the consumers. The answer came from a wealthy young lord, Francis Egerton, the third Duke of Bridgewater.

The Bridgewater Canal

The duke was only twenty-two, and fresh from a visit to the Canal du Midi in France when he decided to have a canal built to carry coal from his collieries at Worsley to Manchester for all the spinning mills that were being built at the time. 'They reckoned,' said Fred, 'the Duke of Bridgewater was thwarted in love so he spent all his energies on this grand plan to get coal from his mines near Worsley to Manchester for all the spinning wheels that were starting to appear there. The Bridgewater Canal was his baby, but, of course, there were some interesting civil engineering things on it, and he employed a very clever engineer to do the work called James Brindley.'

James Brindley

Born in 1716 into a family of yeomen farmers in Derbyshire, Brindley received little formal

education. Instead he was taught to read and write at home by his mother. With her encouragement he was apprenticed to a millwright when he was seventeen and showed exceptional ability. Once he'd completed his apprenticeship, he set up in business for himself and established a reputation for ingenuity and skill at repairing different kinds of machinery. He was soon tackling major engineering projects including the design of a drainage system for the Wet Earth Colliery at Clifton in the Irwell Valley between Salford and Bolton. The mine had been flooded and Brindley's job was to drain it so that it could re-open. His solution was to build a long millrace, 800 yards of which was underground, and took this via a siphon under the river to a waterwheel which operated a pump to clear the mine. It was this ingenious solution that brought him to the attention of the Duke of Bridgewater and Brindley was commissioned by the duke to design the Bridgewater Canal.

Work on the Bridgewater Canal started in 1759 and Brindley provided brilliant solutions to the problems of canal engineering. One of his greatest feats was to build a three-arched stone aqueduct which carried boats forty feet above the river Irwell. Astonished sightseers saw water carried over water and it was so amazing in its time that it was considered a new wonder of the world. You can only see one of its arches today helping to support the Barton Swing Aqueduct. This was built here in the 1890s to take the Bridgewater Canal over the Manchester Ship Canal and allow shipping to pass through. Fred explained:

> This later aqueduct that replaced Brindley's is a very interesting bit of engineering and one that I've always

Fred's eighth wonder of the world, the Barton Swing Aqueduct, carries the Bridgewater Canal over the Manchester Ship Canal

found exciting. When I were a little lad me dad used to say come on today we're going to go somewhere and have a look at the eighth wonder of the world, Barton Bridge. We used to bike all the way from Bolton just to watch it in action. In them days of course it were quite busy. It moved with monotonous regularity. When the canal were first being built W.G. Armstrong and company, the great Newcastle-based engineering firm, got the contract for all the hydraulics which worked all the locks and the bridges all the way along the canal from Liverpool to Manchester. But this bit at Barton is the most impressive, where 800 tons of water in the Bridgewater Canal are supported on a central pivot in the middle of the Manchester Ship Canal. The thing is that when a big ocean-going ship came along the Manchester Ship Canal the swing aqueduct would turn through 90 degrees to allow the ship to get through. Then it turned back to allow barges and smaller boats on the Bridgewater Canal to sail across the top of the Manchester Ship Canal.

If you want to see what Brindley's original aqueduct was like there's another one just a few miles away that's very similar. This one is in the Irwell Valley and it takes the Manchester-Bolton-Bury Canal over the River Irwell. It's incredible really when you think they were in the seventeen odds, and they were taking canals across the tops of great rivers. There were a few of these old aqueducts around Bolton and I remember when we were quite small boys we called them the wooden bottoms because they were actually lined with timber in the bottom and you couldn't sink in the mud in the bottom when you went swimming in it in the summer like you did in the rest of the canal. When they closed the stretch of the canal that went right into the centre of Bolton, there were one of these aqueducts just near where I lived and they actually blew the thing up. They had a tough time doing it but it did give me a really good insight into how the thing were constructed. When they chiselled it apart to blow the arches up, they came across these unbelievable pieces of timber, great baulks of wood about two foot square and about ninety feet long and all completely encased in clay. And when they uncovered it, it were almost like brand new wood, it had been so well preserved in the clay.

When he was building the Bridgewater Canal, Brindley worked closely with the Duke's mining engineer and land agent, John Gilbert, who saw that it was possible to connect the canal directly to his mines by way of a network of underground canals.

The building of Britain's canal network in the eighteenth and nineteenth centuries left us with a legacy of great engineering projects

This in turn could be used to help with the draining of the mines and, at the same time, provide a source of water for the canal. Two canal tunnels were driven into the rock face at the Worsley Delph and a herringbone of minor canals linked these to dozens of pits on either side. Today the Worsley canal basin is hidden away near a busy motorway junction. Tens of thousands of drivers rush by each day unaware of its existence, yet on this spot history was made. The basin with its murky orange water is semi-derelict, but look carefully through the fence and the locked gates that surround it and the remains of a sluice gate and a little wooden boat can just be made out. The water is contained at one end of the basin by a small rock face and just above the water line the tops of two tunnels can be seen disappearing into the hillside. There's little sign of life there now, but two hundred and fifty years ago this basin would have been a hive of activity. Little boats like that one whose rotting hull sticks out of the water would emerge from one of the tunnels, loaded with coal, and then in this basin it would be offloaded into bigger canal boats to be transported to Manchester. Here in its first mile the Bridgewater Canal went deep into the underground workings it had been designed to serve, passing beside underground quays, lit by candles. There miners loaded coal into barges accompanied by the grating, clanking noises of pumps and engines and the sound of muffled explosions deep underground as miners used gunpowder to blast their way into the ground. 'The thing is,' said Fred, 'all the yellow ochre in the water all comes out of the coal measures and the iron ore and that's why all the water's stained orange.'

The first section of the Bridgewater Canal from the Worsley Delph to Manchester was ten and half miles in length and cost nearly fifty thousand pounds to build. It was opened in 1765 and was an immediate success, enabling the duke to cut the cost of his coal by half. 'His enterprise was doubly rewarded,' said Fred, 'because not only did he make a lot of money from it, he also solved the problems of colliery drainage. The water from the mines would come out through the underground canal system. It was brilliant.' Today the quiet waters of the Bridgewater Canal at Worsley give us no idea of what a great engineering achievement it was when it was built in the eighteenth century, or how it revolutionized Britain as it made the transport of heavy goods ten times faster and ten times more efficient than it had ever been before. Before the canals a packhorse could carry no more than 200 lb; a horse pulling a barge on a canal could draw as much as 30 tons.

Canal Construction

'The great success of the Bridgewater Canal inspired even more ambitious programmes of canal-building,' said Fred, 'and Brindley had a hand in nearly every one of them. They say he was a very argumentative man. He were also a brilliant engineer and one of his chief gifts were his ability to have a feeling for the lie of the land just by looking at it. He knew exactly where he were going to put the canal and what its line would be.'

It was the start of a transport revolution that was to change the landscape of Britain. Canals were built to link the new centres of industry with the markets they were serving. The building of Britain's canal network in the eighteenth and early nineteenth centuries left us with a legacy of great engineering projects. As well as needing aqueducts to get across valleys, they also had to get up and down hills. If they came to a small hill they would just take a slice out of it by digging a cutting. If they met a longer hill with a plateau on the top of it, they would build locks to go up one side then down over the other side. And if they came to a big hill they had to build a tunnel to go through it. So the construction of a canal involved many operations other than digging. Quarrying, brick-making, ironwork and joinery were needed for building the stone walls that lined the canals, and constructing locks, bridges, gates, arches and tunnels. Fred's drawing shows the basic principles of canal-building:

> Building a canal like this were a major engineering achievement. There was quite a lot of work involved that nobody could see. All along each bank there were walls that contained the canal made of beautiful stonework. You can see about 18 inches of this above the waterline, but beneath the surface there would be another 5 feet down

Fred's drawing shows the way a canal was built

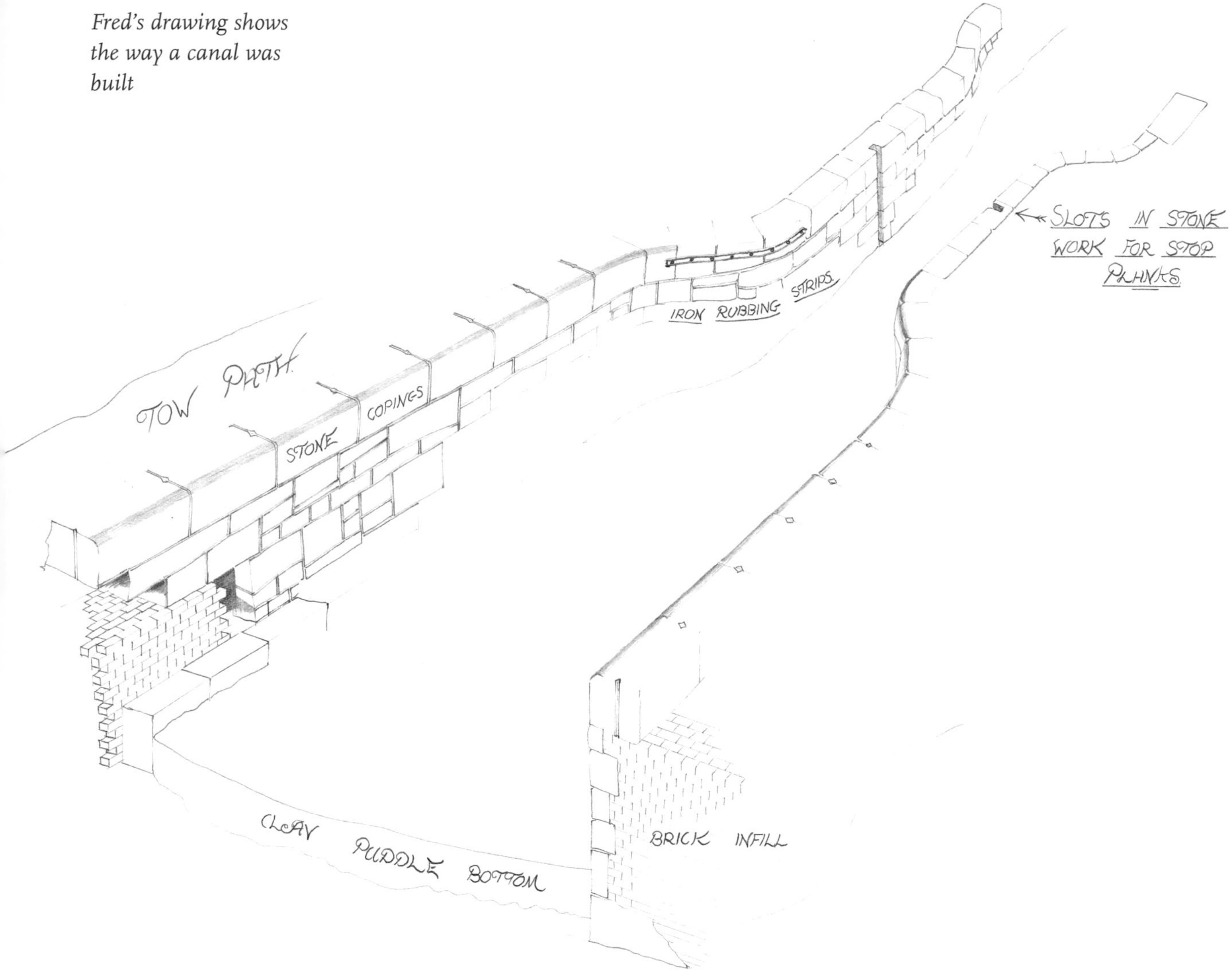

to the bottom of the canal. Behind the actual facing stonework there would be quite a lot of brickwork that were there to give it bulk and weight. Then along the top they nearly always had great big stones which would give the actual edge of it a nice finish. When they were building the canal they lined the bottom with clay to stop the water running out. Putting this clay lining in was known as 'puddling' and there'd be maybe 18 inches or 2 foot of puddle in the bottom.

Puddling is a wonderful canal word for the process of making wet mud waterproof. The commonest and cheapest material used to stop canals leaking is clay, but to work properly that clay has to be mixed with water first, until it forms the tough raw material

It's incredible really when you think that they were in the seventeen odds and were taking canals across the top of great rivers

like that used by a potter. Brindley's method of puddling clay and his use of subterranean channels and aqueducts established him as pre-eminent in the field of canal design and engineering, but like all pioneers he didn't always find it easy convincing the powers-that-be that his way of doing things was the right way, as Fred explained:

> Mr Brindley had a lot of trouble convincing the men of power and Parliament that you could actually dig a man-made river. They all thought that if you dug a trench in the floor all the water would run out of it. Brindley understood the technique perfectly, but in 1762 Parliament demanded more practical proof before they would pass an act to extend the Bridgewater Canal to Runcorn. Apparently Brindley went down to Parliament with a big dollop of clay, stuck it on a parliamentary committee table where he kneaded the clay and water together, made a big hole in the middle of it and poured a bucket of water into it to demonstrate that puddling would make his canals watertight. And of course he won; he got his way, and the canal acts were passed and lots of canals were built all over England.

Brindley was passionate about canals. When a member of a parliamentary committee asked Brindley why God had created rivers his reply was, 'To feed the canals, of course.'

But puddling enough clay for a tea set or for a parliamentary demonstration is a bit different to churning up the many thousands of tons needed to seal hundreds of miles of artificial waterway. It can be done with buckets and spades and the boots of navvies, but in some places mass production was provided by herds of cattle driven along the canal bed as the water was let in, their cloven hooves treading the clay and water together to the required elasticity: the puddle.

The Leeds and Liverpool Canal

By the 1770s canal mania was sweeping the country and plans were well ahead for canal systems linking the Mersey with the rivers Trent, Severn and Thames. One of the most ambitious projects was a canal across the Pennines from Leeds to Liverpool. The original plans were drawn up by a young engineer from Halifax called John Longbotham, whose first idea was to propose a system of canals that stretched from Leeds to Preston. They would carry limestone to the Yorkshire mills and coal to the factories of Lancashire, and the easy access they would provide would slash the costs of industry in the area.

Longbotham's proposal caught the imagination of the merchants in Liverpool,

A canal lock is used to raise or lower the water level

who needed a cheap supply of coal for their factories. So they lobbied hard to have the route changed, to run from the Wigan coalfields to Liverpool. At first they weren't successful, since the original line of the canal agreed by Act of Parliament curved north, bypassing Wigan, and the Liverpool merchants pulled out of the scheme. But without their money, the Yorkshire investors could not afford to build it. They were forced to do a deal linking the canal to the river Douglas, which ran up to Wigan, and the Liverpool merchants got their way. Longbotham's plan had been seen and approved by Brindley who got the job as chief engineer on the grand salary of four hundred pounds per annum. But by this time he'd become involved with a further 363 other canal projects of one sort or another. Fred took up the story:

> I think his workload proved too great for him, because two years later he died. So Longbotham got the job, and what a huge undertaking it was because this was no ordinary canal. The Leeds and Liverpool Canal stretched for 127 miles and climbed over the Pennine chain, the backbone of England. It would link Yorkshire with

Mr Brindley had a lot of trouble convincing the men of power and parliament that you could actually dig a man-made river

Lancashire, and it would compete so successfully with the railways that it nearly drove them out of business. The Leeds–Liverpool was the first of the Trans-Pennine canals to be started but it was the last to be completed. The length and complexity of the route and the obstacles that had to be overcome meant that the canal took forty-six years to build at a cost of five times the original budget.

The canal was constructed simultaneously from both ends and would meet somewhere in the middle. Digging a canal like this was a major engineering achievement. At any one time, the Leeds–Liverpool canal company had between two hundred and five hundred men employed on its construction. There were miners, diggers, quarrymen, masons, carpenters, blacksmiths, lumberjacks and horse-wranglers along with entire teams of surveyors and location scouts mapping out the route ahead and negotiating the deals for the land across which the canal would be built. So the whole enterprise was incredibly expensive, and it was only worth doing if it could dramatically cut transport costs in the area. Building it, however, was also very dangerous. The records of the canal company are filled with names of men who were injured and were paid compensation. From the old canal records book Fred found out about George Clark and Hugh Fraser who received one guinea each when scaffolding collapsed on them in a tunnel they were working on and the company paid the surgeon's bill.

The canal immediately proved its worth when the first stretch running from Bingley to Skipton was completed. On Thursday 3 April 1773, amidst great celebration, two barges arrived at Skipton. They were loaded with coal, which was sold for half the price it had previously cost, a fact that greatly impressed the local newspaper. With the completion of this first section, the landscape began to change dramatically. From Skipton, the route of the canal winds through some very hilly terrain. Naturally, water won't flow uphill, so the canal engineers had to come up with a way of making a long flat stretch of water go up and down hill. The answer was locks.

Locks

A canal lock is used to raise or lower the water level. When a canal boat is going upstream the bottom gate of the lock opens to allow the boat to enter and the paddles on this gate (which allow water to empty) are shut. The gate is closed and the paddles on the upper gate are opened to allow water to flow into the gates lock. These paddles are situated on both sides of the gates and the water is

An eighteenth century civil-engineering masterpiece – the Bingley Five-Rise is the steepest flight of locks in Britain

channelled through tunnels into the lock. As the lock fills with water, the boat begins to rise. When the water is level with the next stretch of canal, the upper gate can be opened and the boat continues on its journey.

The Bingley Five-Rise

There's nowhere better to see the way locks work than the five-rise locks at Bingley. From Bingley, the canal had to climb 120 feet in less than 10 miles, on to the Skipton plateau. So Longbotham used a technique of riser or staircase locks, perfected by his hero Brindley, which allowed the canal to rise much more steeply than usual. The lock staircase that he constructed is an eighteenth-century engineering masterpiece and the most spectacular feature of the canal. The Five-Rise consists of five locks connected together. They open directly one to another, with the top gate of one forming the bottom gate of the next. This unique staircase is the steepest flight of locks in Britain and has a total fall of 60 feet. When the locks were completed in 1774, thousands gathered to watch the first boats make the descent. Now, over two hundred years later, the flight is still in daily use. The staircase is slow to operate and all five locks must be 'set' before beginning the passage. Then it takes just over half an hour for a boat to go up or down the lock sequence. For a journey upwards the bottom lock must be empty with all the others full while for a boat coming down the flight the top lock must be full and the others empty. The Leeds–Liverpool Canal has many lock staircases of two or three locks each, including one only a few hundred yards downstream from the Five-Rise – this one a three-rise flight with a fall of 30 feet. Over on the other side of the Pennines, the Wigan flight lifts the canal over 200 feet in the space of a mile.

Because this canal is wide each lock gate consists of two half gates, hinged from opposite sides of the canal with the two halves closing in a V shape pointing upstream. This means that water pressure on the uphill side of the gate keeps it tightly closed until the water levels on either side are equal, when the gate can be opened and the boat in the lock can be moved to the next chamber. Lock gates may be just big doors, but because they have to control huge water pressures and take the hard usage they get from boats barging into them, they are constructed with massive strength. To be waterproof they also have to be built very precisely, fitting tightly against the masonry of the lock walls and to each other. All this structure then has to be carefully balanced and pivoted so that it can be opened and shut by one person safely. Then they are left outside and underwater and expected to

survive for a long time. 'There's more to lock gates than meets the eye,' Fred said:

> When you first look at them you think they're just a great pair of waterproof doors. But they're not really doors because they have no hinges; they are almost floating. They make them out of great lumps of wood and there's a great lump of timber sticking out at the side. At each side of the lock where the hinges should be there are two semi-circular grooves in the stonework of the lock chamber and they are curved to the shape of the piece of timber that is sticking out of the lock gate. And when the pressure of the water fills the lock up it pushes the two edges of the gate into the grooves and, of course, forms a watertight seal. Later on they tried making lock gates out of iron. I suppose it was a sort of economy measure, but it didn't work because when it is under great pressure, iron bends, and once you've bent it, it doesn't want to come back. Wood, which is of course a bit more expensive, it's a dead cert it'll work. The gates have got to be so tough and strong because of the bashing about they get from the boats that were toing and froing every day. When you think of all the effort it takes to get through just one lock it's quite an achievement to sail the whole length of the canal with another ninety to tackle.

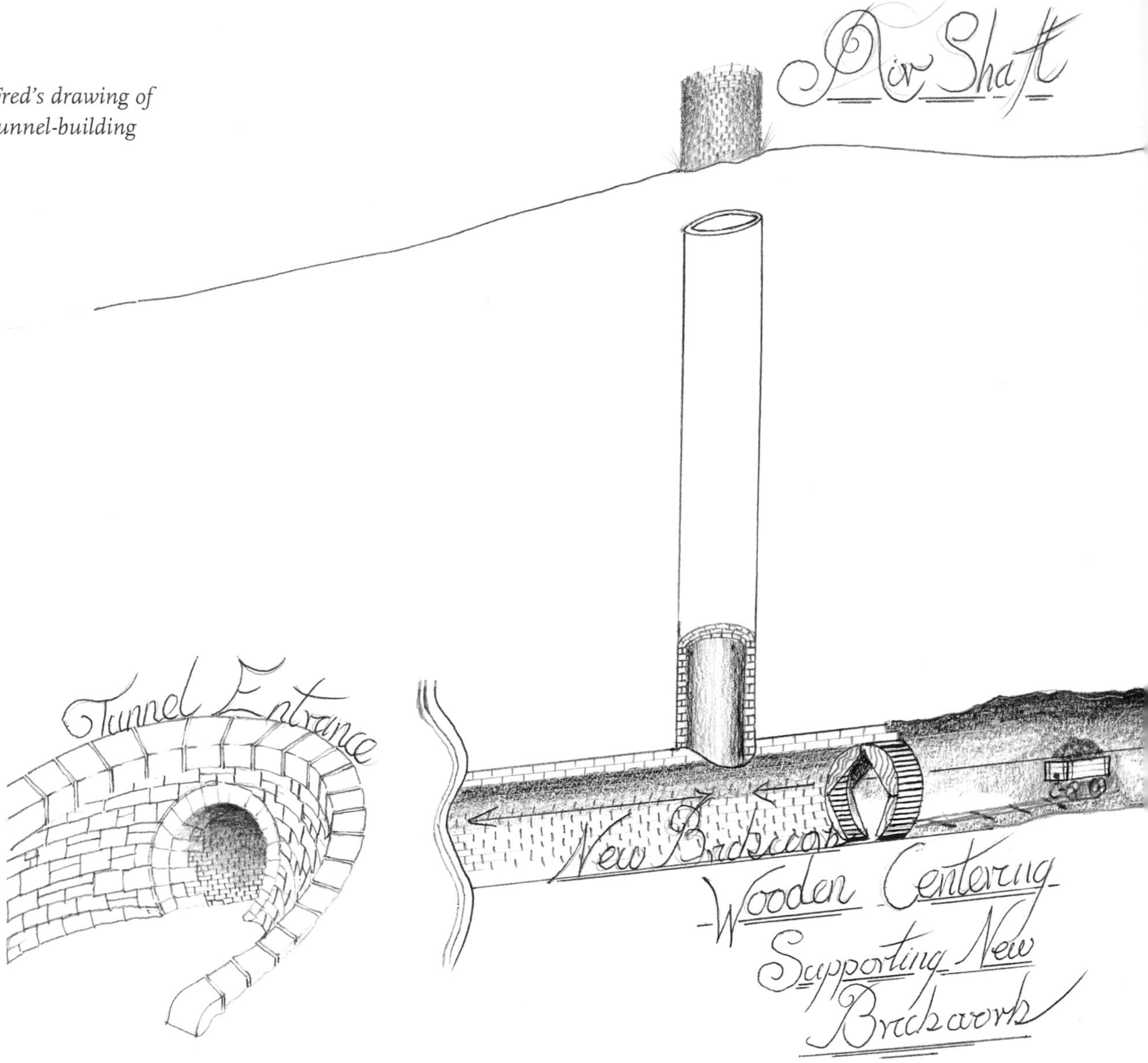

Fred's drawing of tunnel-building

The thing about staircase locks is they use an awful lot of water. Each lock chamber in the Bingley Five-Rise takes 80,000 gallons of water. So every time you use it, at least 80,000 gallons is being poured from the top of the hill to the bottom. All that water has to come from somewhere, and it comes from a series of reservoirs at the top of the canal system. Foulridge Reservoir is actually a series of three reservoirs at the highest point of the canal, which feed water down into the canal. When the water level of the canal drops they are able to get water from the reservoirs down through channels and into the canal which quite literally runs below the reservoirs along a tunnel through the mountains. So there is always water to replenish the canal and keep it working.

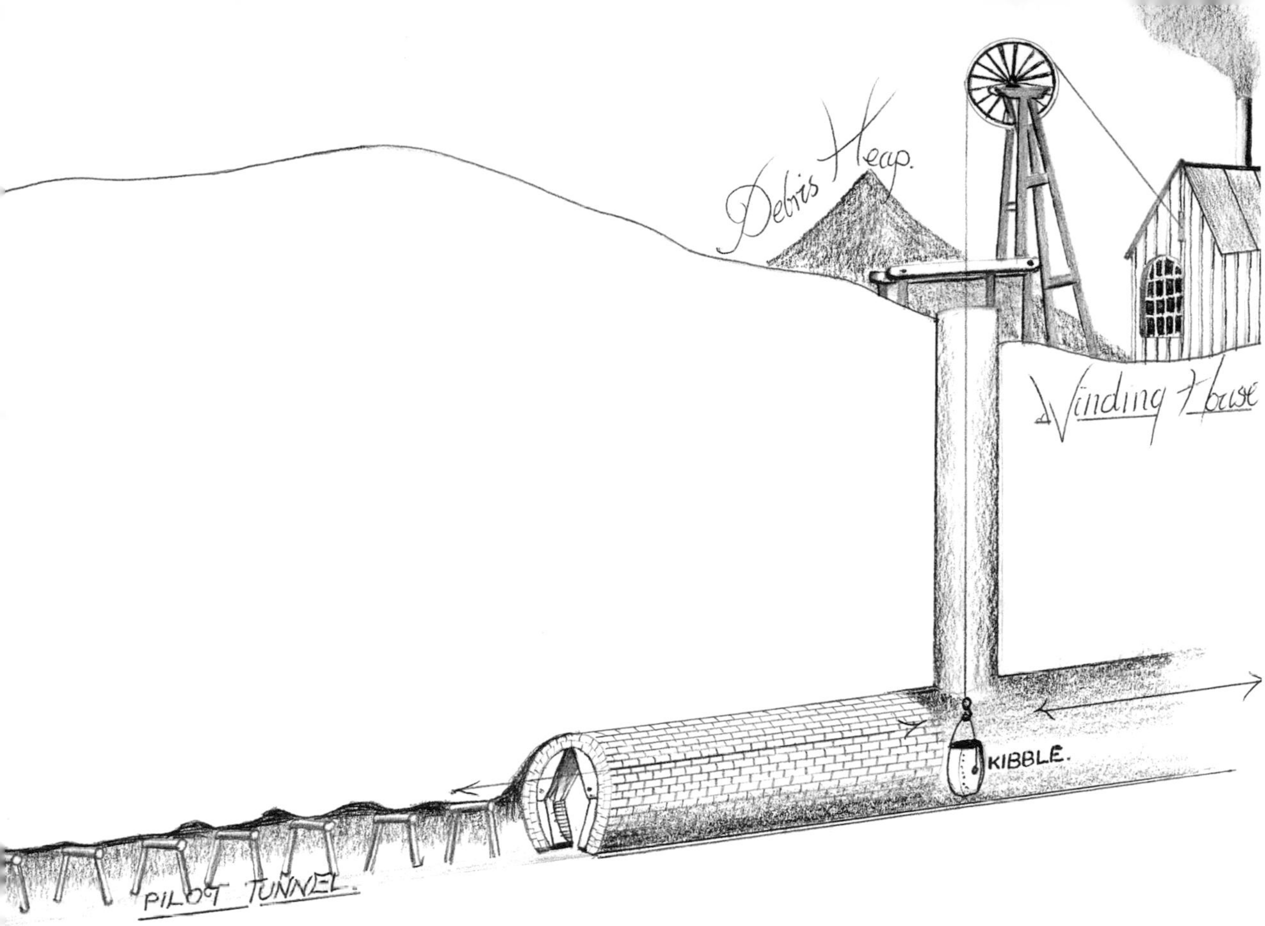

The Foulridge Tunnel

Tunnel-building was probably the most difficult engineering task the early canal-builders were faced with. Surprisingly, for a canal that crosses the Pennines, the Leeds–Liverpool has only one tunnel, at Foulridge near Colne. From Skipton, the canal rises another 143 feet to Foulridge, where a mile-long tunnel was dug through the Pennines. With the aid of one of his drawings Fred gave an account of how a canal tunnel like this would have been built.

Tunnel-Building

Tunnels were usually built by laying a straight route across the hilltop and then sinking a number of shafts. The first thing they would do would be to walk over the top of the hill, going in as straight a line as they could with, I suppose, the equivalent of a telescope or something just as simple as that. They'd mark out a series of pegs and then sink a line down to the level of where the tunnel was going to be. And then they would proceed to drive headings from each end of the bottom of the shaft. They would have a semi-mobile winding gear and the debris would be raised up the shaft in a kibble which is the name for

Fred demonstrates the cut-and-cover method of tunnel building that was employed at Foulridge

a small iron barrel. As the line of the tunnel advanced across the mountain they would dismantle the winding gear and move it up the road to the next shaft and do the same thing again. Once all the shafts had been sunk, digging began in both directions from the bottom of all the shafts and from the tunnel entrances. The early tunnellers were miners who were not used to taking accurate headings, so the horizontal shafts sometimes didn't meet up correctly.

At Foulridge all was going well until they came across a problem they hadn't reckoned with. At each end of the tunnel they found there were shifting sands and they found the lakes and water above were coming in constantly on them and created trouble. It must have been almost impossible, no wonder it took almost six years. Because all of these problems they had to use two different methods of tunnel building.

What they had to do was use what is known as the cut-and-cover method of tunnelling. For this they dug a great cutting through the hillside and then put in the centring. That's the wooden framework that was used in arch and vault construction by the builders of the cathedrals. As in the cathedrals the centring would be made of wood. Once the centring was in position, they would then proceed to lay the masonry on top of it and this had all been cut to shape with the right angle on. Because it was in a cutting and it was all exposed to daylight, they could do a much better job and use much bigger stones. After the last stones had been firmly cemented into position, they would then proceed to cover the whole lot up – very carefully, I should imagine. They wouldn't have really chucked it around but would have been making sure that the pressure as they filled it in were equal on both sides to squeeze the arch down on to the centring. This is very important and there were lots of disasters when this was not quite right.

After they'd got the masonry and the centring all buried under thirty feet of unstable ground again, they would then proceed to withdraw all the wedges out from underneath the centring which of course would have the effect of the centring being lowered and then they could withdraw it. So they could just move it up a bit and put some more masonry on and keep advancing like that until they'd gone all the way through the hillside as you might say. When the centring's been removed you end up with that beautiful smooth stone arch tunnel. When they were building these tunnels they must have moved literally thousands of tons of dirt with wheelbarrows and no doubt they would be up to their necks in mud as well while they were doing it. It's a credit really to our illustrious ancestors.

Navvies

The canal network laid out in Britain was known as the 'Inland Navigation System'. The men who dug the canals, officially called Excavators, became known colloquially as Navigators, which was later abbreviated to Navvies. Fred described them:

The men who actually built the canals and did all the tunnelling were professional navvies who travelled along the line of the canal in vast encampments. The locals viewed them as lawless gypsies. Their equipment was primitive. All the digging was done with pick and shovel and 'spoil' [soil] was moved by hand in small barrows. The navvies were frequently undisciplined. They lived in squalid hutment townships and spread terror to the surrounding villages, especially on pay nights. In the canal company records, I found an account of a serious riot that happened at a place called Barrowford in 1792. It says a riot of a very serious nature occurred amongst the town people of Barrowford and the workmen employed on the canal. Several workmen were injured and the fighting had to be broken up by the local militia. The canal committee issued five hundred notices threatening to prosecute anybody who incited any more violence.

Once the Foulridge Tunnel was completed in 1776, a crew could leg a boat through the tunnel in just over $1^1/_2$ hours. Legging was the common method for getting a barge through a canal tunnel, Fred explained:

When the canals were in operation the way they got through the tunnels was interesting. Modern boats have got engines, but of course in the olden days they had horses. The tunnels didn't have

Burnley embankment from which you can look down over the town

towpaths running through them, so what did they do with the horse when they came to a tunnel? Well, it's quite simple really. They used to let the horse walk over the top of the hill by itself or one of the boat crew would lead it over, one of the kids maybe, and then they'd have to use manpower to get the boats through the tunnel. There'd probably be about 30 ton of goods in the boat and the boat would probably weigh about 15 tons. One method to get the barge through the tunnel was to use a boat shaft and push along the roof of the tunnel and walk along the top of the cargo, but that used to wear the bricks away in the roof of the tunnel, so the canal company owners preferred them to use the art of legging. Two men would lie on the roof of the barge with their legs up on the roof of the tunnel and they would literally walk the loaded barge through the tunnel.

By 1796, after the Foulridge Tunnel had been completed, the Leeds–Liverpool Canal stretched all the way from Leeds to Burnley. It crossed the Burnley plain on a mile-long embankment, elevated above the town. From a barge you feel as if you are sailing above the rooftops. At the Liverpool end, though, it wasn't such plain sailing. Here they faced stiff opposition from a rival project called the Lancaster Canal. This was a scheme to run a canal south from Kendal, through Preston and down to Worsley, where it would link with the Bridgewater Canal. The only problem was, its planned route ran slap-bang across the line proposed by the Leeds and Liverpool. Both sides raced to get their

The early tunnellers were miners who were not used to taking accurate headings, so the horizontal shafts sometimes didn't meet up correctly

canal's Bill passed through Parliament, and the Lancaster Canal won. The Leeds and Liverpool was stopped in its tracks. The Lancaster Canal Company dug a canal from Wigan towards Preston, but at Johnson's Hillock near Chorley they ran out of money. The Leeds and Liverpool proprietors seized their chance. They did a deal with the Lancaster, merging the two canals and digging a sequence of twenty-one locks down the hill at Wigan, to link the bottom end of the Lancaster Canal with the eastern section of the Leeds and Liverpool. Very cannily, they limited the length of these locks to 62 feet, allowing only Leeds and Liverpool boats to pass through them, rather than the 74-foot-long boats of the southern waterways. This effectively gave the Leeds and Liverpool company a stranglehold on canal traffic north of Wigan. The two canals merge at Johnson's Hillock, where on Tuesday 22 October 1816, members of the Leeds and Liverpool Company travelled from Leeds on board the *John Hustler*, and were joined by members of the Lancaster Committee. Together, they symbolically travelled on to Liverpool, where they arrived at five o'clock on Wednesday amid tremendous celebration. Now, the Leeds and Liverpool Canal had truly been built.

Altogether it had taken nearly half a century to complete the link all the way across from Leeds to Liverpool. Boats were now able to ply their trade all the way across the Pennines, carrying wool from Leeds, coal to Liverpool and limestone from Skipton all along the length of the canal. The cost of raw materials was slashed as the costs of transport came tumbling down. The Leeds and Liverpool Canal was a great commercial success. Each barge which used it had to pay tolls along various stages of the route at toll houses. At just over 127 miles the Leeds and Liverpool is the longest man-made canal in Britain and the civil engineering that went into its construction is awesome. The Five-Rise staircase locks at Bingley and the 1,000-yard-long embankment at Burnley are both classed within the seven wonders of the UK's Inland Waterway network. Few other canals in the country carried such a large range of commodities or served such a diversity of industry. The result of this was that it had a major effect on the growth of the communities through which it passed.

The canal network changed the face of Britain and cities like Leeds, Manchester and Birmingham grew and prospered on the basis of this new transport system. But out of the Industrial Revolution, which covered the country with mills, factories and smoke-filled industrial towns and which the canals had done so much to create, came the invention which superseded them: the railway.

Railways

With the coming of the railways the majority of the canal network that had spread across Britain so rapidly was abandoned. It was by appreciating and overcoming the engineering difficulties involved in canal-building that the civil-engineering profession came into being. The construction methods pioneered by the canal-builders were adapted by the railway engineers of the nineteenth century and the labour force on these projects continued to be known as navvies.

In the nineteenth century, armies of navvies were employed in building a railway network that covered Britain

The canals were built between 1745 and 1830, by which time there were almost 4,000 miles of navigable waterways throughout Britain. With the opening of the railway linking Liverpool and Manchester in 1830, the railway age began in earnest. Railway-building assumed a frantic pace. Soon armies of navvies were employed building a railway network that covered Britain. At the peak of 'Railway Mania' in 1847 there were 250,000 of them at work all over the country. The railway network grew at an amazing rate. From less than 100 miles in 1830 there were over 1,000 miles by 1837 and within twenty years of the opening of the Liverpool and Manchester Railway more than 5,000 miles of track covered the country. By the turn of the century the navvies had built almost 20,000 miles of railway.

Brunel's Box Tunnel – nearly two miles long, it was, in the 1830s, the longest railway tunnel anybody had ever built

The three bridges over the River Conwy – Stephenson's tubular railway bridge is on the right, Telford's suspension bridge in the centre

Robert Stephenson

In the early years of Queen Victoria's reign, George Stephenson's son Robert was in the forefront of creating a railway network which was to transform the lives of millions. After serving an apprenticeship at Killingworth colliery, he helped his father survey the Stockton and Darlington Railway. Then in 1823 he became a managing partner in Robert Stephenson & Co., where *Rocket* was built. Throughout the 1830s the company began to build increasingly powerful locomotives and his design for a long-boilered mainline engine was still in use in the 1890s. An engine based on this design can be seen at the National Railway Museum and it illustrates the rapid advances that were made in a very short time from *Rocket*. But Robert Stephenson and his company didn't just build the locomotives; they built the lines and the bridges and all the engineering works involved in the construction of a full-size railway.

In 1833 Robert became chief engineer of the London and Birmingham Railway. After battling with engineering, political and commercial problems, he completed the construction of the 112-mile long railway in 1838. His greatest accomplishments, however, were his tubular railway bridges over the Menai Straits and the river Conwy, the High Level Bridge in Newcastle upon

I think he kept himself going chain-smoking cigars and working twenty hours a day

Tyne and the Royal Border Bridge at Berwick-on-Tweed. He used a novel box-girder type of construction in his bridges across the Menai Straits and at Conwy in which the railway tracks were completely enclosed in parallel iron tubes. Newcastle upon Tyne's High Level Bridge is quite an unusual structure. 'It's a two-tier job,' said Fred, 'the locomotives are on the top and the road traffic is underneath. It's basically a collection of cast-iron arches held together by wrought-iron tie rods supported by five sandstone pillars. It was opened in 1849 by Queen Victoria and it's still going strong to this day. Really it's a credit to him.'

Robert Stephenson was one of our greatest railway builders and railways had become the symbol of Victorian industrial might. But there was another man who had an even greater impact on the Victorian Age.

Isambard Kingdom Brunel

'The man who was really Stephenson's great friend and one of his greatest rivals were even more innovative,' Fred said, and went on to describe him with evident admiration:

> Throughout his great career he built railways, bridges, tunnels, ships, the lot. That's why Isambard Kingdom Brunel is my hero. The son of a French-born engineer Marc Isambard Brunel, he was born in Portsmouth in 1806 and educated in both England and France. Brunel had unshakeable confidence in himself. He were so confident that he actually invested his own money in most of the projects that he undertook. His energy was immense and there's no detail of any project he were involved in that he didn't want to know everything about. I think he kept himself going chain smoking cigars and working twenty hours a day. He earned and greatly valued the esteem in which the navvies, the shipwrights and the other engineers held him and they actually christened him the Little Giant, because he was only little, you see.

Nearly all Robert Stephenson's work involved railways. Brunel, on the other hand, would have a go at anything. In fact his first major engineering project involved tunnelling. The tunnel under the river Thames between Wapping and Rotherhithe was handed down to Brunel by his father. The tunnel is over four hundred yards long and it was the first tunnel under a river of any great size. In spite of all the great engineering wonders he went on to, I think this tunnel was one of Brunel's greatest achievements. It was his father, Marc, who took on the job of digging a tunnel under the river Thames and I think it was the first tunnel ever to be attempted under water. Marc Brunel started work on the tunnel but

Brunel's Clifton Suspension Bridge spans the 700-foot deep Avon Gorge

he got hurt in an accident. Then he got pleurisy so he went into a sort of semi-retirement and left young Isambard, who was only twenty-one years old, to complete the thing. Isambard sank two great shafts, one on each side of the river, at Rotherhithe and Wapping, and the grand plan was to have a spiral roadway going down the shafts to the bottom. At the bottom of each of the spiral roadways were the entrances to the tunnel itself. It was supposed to be completed in three years, but it didn't work out like that because they kept running out of money and there were arguments with the directors and one or two inundations of water from the river above.

At the time the Thames was not much more than a sewer and where they were digging was not very far below the riverbed. During the building operation there were many occasions when the polluted Thames water came flooding into the tunnel and, although there was no loss of life, the raw sewage that came flooding in had a very bad effect on the eyesight of the workers and on their health in general. One day, though, just as they seemed to be making good progress, the water suddenly came flooding in and washed them all out into the great chamber at the end. Brunel was in the tunnel at the time and, along with all the rest of the workers down there, he was lucky to get away with his life. Everybody did manage to get out, but the pumps couldn't get rid of the river water. To solve the problem they bought a diving bell and sailed out with it to the middle of the river in a boat. When they found the hole, they put a great sheet over it and bags of clay and twigs. They let all of this sink on to the bed of the river and then they started the pumps and they pumped the water out from the tunnel. Then, just like the miners in Cornwall, he plugged the hole in the tunnel roof with clay and puddling and started again. You have to hand it to him, when you think the guy was only twenty-one years old and he was doing a thing like this that nobody else had ever attempted before. I should imagine it would have been a wet and smelly job because the river Thames were almost an open sewer. A lot of the navvies digging the tunnel were affected with disease and problems with their eyes. It was all very dirty and dangerous and when disaster struck again Brunel nearly lost his life. But he'd started to build his reputation.

While he was convalescing in Bristol, Brunel got to know a group of merchants and industrialists. Through them, as Fred explains, he got involved in a daring scheme to bridge the 700-foot-deep Avon Gorge.

Brunel had an ability to sort out committees and boards of directors and convince them about the practical side of his ideas and his big projects. At Clifton when they had the competition to build the bridge he put in two designs and they had no less a man than Thomas Telford to be the judge. The first two designs that he submitted were turned down but Brunel didn't give in with his persuasive powers. He managed to turn round the committee's decision and ended up having his designs for a suspension bridge accepted.

At the beginning of the construction of the bridge at Clifton one of the first things they did was to blacksmith an iron rod 2 inches in diameter and they stretched it from one embankment to the other. I should imagine they first started off with a piece of string and then a bit of decent rope. Onto this they would have attached a wheel with some sort of carrier suspended from it. Then as each rod were hammer-welded on the side of the valley, they would have pulled it across on the carrier.

Unfortunately the directors didn't raise enough revenue to complete the bridge so work on it had to be stopped and it was only finished years later after Brunel's death as a memorial to him. But it was through this same group of Bristol merchants who had commissioned him to build the bridge that he got involved with an ambitious project for a railway between Bristol and London. This was authorized as the Great Western Railway Company in 1835 with the twenty-nine-year-old Brunel as its first engineer.

It was one of the most exciting things we did in all of our filming – to be able to read the actual letters and papers of my greatest hero

The Great Western Railway

Much early railway development was haphazard and the Great Western was no exception: Brunel went his own way by refusing to adopt George Stephenson's 4-foot 8½-inch standard gauge for its tracks. Brunel's vision was one of a complete railway system, which would include innovations in architecture, track work, locomotives and, above all, track gauge. He felt that Stephenson's 4-foot 8½-inch gauge was too narrow so he made his 7 feet 0¼ inch wide. But every other railway in the country standardized on the 4-foot 8½-inch gauge so it was inevitable that Brunel would lose the battle. The tracks we now travel on are Stephenson's narrower gauge.

By 1838 the line was open from London to Maidenhead, where the railway crossed the Thames over the longest and flattest brick arches ever built. With a length of 128 feet and a height of 24 feet it is a record that still stands today. From London to Chippenham the line was fairly flat and they called it Brunel's billiard table, but when he got to the section between Chippenham and Bath it was full of obstacles. It involved many deep cuttings and a tunnel at Box. Nearly two miles long, it was by far the greatest railway tunnel anybody had ever attempted, as Fred described:

It was a huge undertaking, especially when we remember that apart from the steam pumps which kept the workings clear of water and the power of gunpowder that was used to blast away the rock, it was all done entirely by the strength of men and horses and the whole proceedings were lit by candle power. I mean, when you think about it, what an achievement!

To find out more about the sort of man Brunel was I went to Bristol University Library where they have a collection of his letters and diaries. It was one of the most exciting things we did in all of our filming; to be able to read the actual letters and papers of my greatest hero. From reading some of them he was obviously a man of great ambition and drive and he spent lots of time away from home and his family, but he did keep in touch and wrote back home from time to time. There's a lovely letter in the library to his wife Mary that shows what sort of a guy he was. In it he says, 'I have walked today 18 miles from Bathford Bridge and I am not really tired.' In fact he goes a bit further on and says that if he'd got there a bit earlier he would have caught the train down to London and come back on the goods train early in the morning. What a fella; it's harder than climbing chimneys that! He says, 'The hotel which is the best of a deplorable set of public houses is full,' and therefore he was staying at the Cow and Candle Snuffers and he goes on to describe his room. 'There are four doors and two windows. What's the use of the doors I can't conceive for you might crawl under them the gap's that big.' Then he ends his letter, 'Goodbye, my dearest love, yours, I. K. Brunel.

By June 1841 the whole line from London to Bristol was open, and Brunel started work on an extension to Exeter which would span the river Tamar at Saltash. The bridge at Saltash is one of his masterpieces. It's the only railway-carrying suspension bridge in the country and a magnificent engineering achievement. His confidence in taking on projects like this was amazing.

In the 1920s Great Western Kings and Castles became a benchmark for locomotive design

The Heroic Age of Engineering

The Forth Bridges – demonstrating the cantilever design of the railway bridge and the first crossing of the road bridge on a tractor engine

During the Victorian era engineers and their locomotives were virtually unstoppable, constructing a bridge or tunnel or cutting an embankment at every obstacle. As the nineteenth century progressed, bridge-building became more daring and dramatic.

With a length of over two miles the Tay Bridge is the longest railway bridge in Britain, but it is its predecessor that is notorious in the history of British civil engineering. The 'Beautiful Railway Bridge of the Silver Tay', designed by the engineer Thomas Bouch, was opened in 1878. But just one year later it collapsed in a storm just as a train was crossing it. The engine and all its carriages plummeted into the river below and all seventy-five people on board lost their lives. It was one of the worst disasters in British railway history and an enquiry concluded that it was caused by faults in the design, construction and maintenance of the bridge.

The Forth Railway Bridge

Prior to the collapse of the Tay Bridge, Bouch had put plans in for the construction of a railway bridge over the river Forth, but the disaster and resulting loss of confidence in his expertise put paid to that. Instead a cantilever design was put forward by John Fowler and

Blackpool Tower – a wonder of Victorian engineering

Benjamin Baker. Their plan – to build the biggest railway bridge in the world not with iron, but with the relatively new material of steel – was accepted. The basic engineering principle of the cantilever was well established – a horizontal projection like a balcony or canopy supported by a downward force behind a fulcrum. In their design for the Forth Bridge the main crossing consists of tubular struts and lattice girder ties in three double cantilevers each connected by suspended girder spans resting on the cantilever. Each of the 361-foot-high double cantilevers is supported on granite-faced piers sunk into the bed of the Forth. The Forth Bridge was the first bridge to use the new steel produced by the Siemens open-hearth process and it was designed using the most thorough structural analysis possible. The bridge is an enormous cantilever construction with 1,700-foot main spans: a world record for many years. Its three double cantilevers are known as the Queensferry, Inch Garvie and Fife cantilevers. This colossus, the like of which had never been seen before, rose above the waters of the Firth of Forth between 1883 and 1890. 'But,' said Fred, 'it is my opinion that the Tay Bridge disaster is the reason why the Forth Bridge is so grossly over-engineered.'

The need for the Forth Bridge was quite simple. The North British Railway Company, whose main line between Edinburgh and Dundee was cut by the Firths of Forth and Tay, was under pressure to bridge the gaps. Rail links across both of the estuaries had been seen as inevitable for a generation; the only things that were needed to make it happen were the technology and the materials. The technology was acquired painfully with the disaster on the Tay and the material, steel, had become available in quantity thanks to Bessemer's recently developed process.

In 1882 a contract was placed with Tancred Arrol & Co., a consortium led by William

The Forth Railway Bridge was the largest civil engineering structure completed during the nineteenth century

Arrol of Glasgow, who was contracted to build the second Tay Bridge also. The riveting machines used on the Forth Bridge were of his own design, developed specifically to overcome the unique problems of building this huge structure. He went to work in a cotton mill at the age of nine and at fourteen he became a blacksmith's apprentice. Eventually he set up his own business, establishing the Dalmarnock Works, and he undertook his first major construction contract, the North British Railway Bridge across the Clyde at Bothwell. While working on the Forth Bridge and the Tay Bridge he was also working on London's Tower Bridge. Later contracts included the Nile Bridge at Cairo and the Wear Bridge at Sunderland. The engineers for the project were Benjamin Baker and John Fowler. Baker had a flair for engineering design with an understanding for materials gained while serving his apprenticeship at the South Wales Ironworks at Neath Abbey. He moved to London after his training where he worked on the construction of Victoria Station and much of the London Underground Railway Network. John Fowler, the consulting engineer, was considered one of the best of his time too. He designed the Pimlico Bridge, which carried the first railway across the Thames in 1860, and was a pioneering engineer of the London Underground system. He also worked on St Enoch's Station in Glasgow. On the death of Brunel he became consulting engineer to the Great Western Railway.

When it was opened in 1890, the Forth Bridge was the largest that had ever been built and the first with a steel superstructure. The bridge, including the approach viaduct, is 2,765 yards long, $1^{1}/_{5}$ miles. The actual length of the cantilever portion in the middle is 1 mile 20 yards. In its construction over 50,000 tons of steel were used and the height of the steel structure above mean water level is over 370 feet, while the rail level above high water is 156 feet. About 8 million rivets were used in the bridge and 42 miles of bent plates used in the tubes. This was the largest civil engineering structure completed during the nineteenth century. When the bridge was completed it was the largest span bridge in the world. Thousands of men from all over Europe laboured day and night for seven years, and at the peak time of its construction 4,000 men were employed. It was made in sections on the shore at Queensferry, where all the rivets were tested before they were shipped out to site. In effect, the bridge was built twice, such was the obsessive attention to detail and safety.

Engineering on this scale didn't come without its price and 57 lives were lost during the construction of the bridge; 518 men were

When you stand on the very top of it, 350 feet up in the sky and a locomotive comes onto the bridge you can feel the whole thing rock

taken to hospital while the fate of 461 injured men is not recorded. However, this toll seemed relatively light when the dangers of the project were taken into consideration. The men were working at great heights, but the wire safety ropes they used saved many lives, accidents could have been doubled without them. But they were also working below the waterline. To provide the foundations for the piers, caissons had to be dug deep below it. They were a form of air chamber, resembling a well, which were driven to a firm foundation stratum under the bed of the river and filled with concrete. That there were hardly any accidents during the caisson operations was a remarkable achievement. The men on the caissons quite literally dug away at the foundations on which they stood. If the caisson sank too quickly, they could all too easily be squashed between the seabed below and the thousands of tons of concrete immediately above them. On one occasion several men were buried up to their chins in the mud, and on another, a caisson suddenly dropped 7 feet while men were working underneath. The Queensferry north-west caisson ruptured while being pumped out. The sea rushed in and two men were killed.

The bridge was considered to be a triumph of engineering skill and, at the time, was said to be the eighth wonder of the world. It would, it was predicted, lead to a revolution in the art of constructing bridges of this kind. The grand opening was on 4 March 1890 when the Prince of Wales drove in the last rivet. The Victorian public finally had what they wanted: safety, reassurance, stability and strength and the achievement put Scotland into the forefront of civil-engineering expertise. Fred described the bridge:

> It's now a hundred and ten years old and a major refurbishment is under way which gave me the chance to have a good look at it. It's amazing when you think how the great cantilevers are not really mechanically connected at all. In order to allow for contraction and expansion they are just linked up together like a chain. It is because of this, of course, that when you stand on the very top of it, 350 feet up in the sky, where we were, and a locomotive comes on to the bridge under the cantilevers, you can feel the whole thing rock. Quite a fantastic feeling and a credit to the men who built it. And all based on the cantilever principle. I would have loved to have seen it when steam trains used to come thundering across and to have been able to get up there on the girders with one of the painting gangs. It all looks very exciting. Today the bridge is still very busy. It carries 150 trains a day, but most of them are just little diesels.

The great Gothic-towered, hydraulically operated bridge has been a symbol of London since 1894

Tower Bridge

Tower Bridge is another great Victorian engineering feat built at the end of the nineteenth century and we went to London to see this world-famous landmark. The great Gothic-towered, hydraulically opening road bridge next to the Tower of London has been a symbol of London since 1894. Inside its castle-like exterior there's an enormous steel frame that was constructed by the same men who built the Forth Bridge. Altogether there're about 11,000 tons of steel in the towers, the roadways and the walkways. 'When you come into one of the towers,' Fred said, 'you can see its great steel skeleton that's all riveted together. Beautiful riveting it is as well. The whole thing would stand up really without the fancy stonework or beautification on the outside. It's a wonderful bit of ironwork really.'

Before the Victorian age there had never been a bridge downstream of London Bridge. But the massive growth in population in the East End made a new one essential. The problem was that this stretch of the river had some of London's busiest docks and any new bridge would have to give up to 140 feet of clearance for tall ships making their way into them. The solution came from Horace Jones, the city architect, with his design for Tower Bridge. It took eight years to build and five major contracting companies were involved, employing between them around five hundred men. On the completion of the steelwork the bridge was clad in Cornish granite and Portland stone, both to protect the ironwork and to give it the beautiful appearance it has now, which makes it tone in with the Tower of London next to it. The

Blackpool's North Pier and Tower helped transform the town into Britain's busiest seaside resort

bridge is hydraulically operated and much of the original machinery can still be seen in the engine rooms, including the steam engines that they used to power the enormous pumping engines. It was manufactured by Lord Armstrong, the owner of Cragside. The energy that was created was stored in six massive accumulators, so that as soon as power was required to lift the bridge it was readily available. 'When you look at it you'd think it was like a drawbridge going up and down. But down below in the foundations there are great curved boxes full of iron to counterbalance the weight of the roadway so there is not much power needed to actually open it up,' Fred explained.

Blackpool Tower

'Not far from me we've got an engineering wonder of a very different kind,' said Fred, 'and like Tower Bridge it's one that's become one of the country's most famous landmarks. For me Blackpool Tower is another wonder of Victorian engineering.' The tower, which dominates this part of the Lancashire coastline, took three years to build and was opened in 1894. Again, it was regarded as one of the greatest pieces of British engineering of the time. It was built in imitation of the Eiffel Tower, which was opened in Paris in 1889, and it soon became a popular attraction, helping to turn Blackpool into Britain's busiest seaside resort. People had first started visiting Blackpool for health reasons. In the late eighteenth century the medicinal values of the seaside were much in fashion and visitors would often drink up to twelve glasses of seawater a day! However, it was the introduction of the railway to Blackpool that saw a massive expansion in the number of visitors. In 1849 a convoy of trains left Bolton for Blackpool with over ten thousand people packed into nearly two hundred carriages. Local traders got together to see how they could take advantage of this dramatic increase in visitors and the result was to build Blackpool's first man-made holiday attraction: North Pier. 'That in itself,' said Fred, 'was a bit of an engineering wonder. It was designed by a gentleman called Eugenius Birch. That's some name, innit? And he decided that he would build it out of cast-iron stanchions instead of the much stronger wrought iron. His argument against the wrought iron were that if a ship crashed into it, it would bend it and buckle it and twist it. If a ship crashed into his cast-iron stanchions it would bust a few and they would be able to replace them. I think that were a good idea myself. I do indeed.'

Opened in 1863, the pier was an immediate success with over a quarter of a

million people paying the 1*d* entrance toll in the first year. The pier company aimed to appeal to a high class of visitor and only two kiosks were allowed on the pier, one selling books and the other tobacco. Soon after Central Pier was built and became the first attraction to cater for the working-class visitor. The 'People's Pier', as it was known, became famous as an open-air dancing venue, where live music would continue into the night, in stark contrast to the much more subdued activities of the middle classes over on the North Pier. The tower was designed to incorporate all of Blackpool's traditional attractions but in more luxurious surroundings. The ballroom appealed to those from the North Pier who liked to dance while the circus below could be filled with water to allow swimming and diving shows that had been popular at the end of North Pier.

It took just under 2,500 tons of steel and 93 tons of cast iron to build the tower and at any given time there were around two hundred men working at great height to finish it. It rises to a height, at the top of the flagstaff, of 518 feet with an observation platform that is 480 feet above the base, and it helped transform Blackpool into one of the biggest and busiest tourist resorts in all of England. Fred pointed out:

> What you've got to remember is when it was opened in 1894 there were no aeroplanes and no skyscrapers and most Victorian people had never been very high

To build it, I'm pretty sure that they would have employed steam power

off the ground. To actually have the experience of being 500 feet up in the sky and being able to see 30 miles must have been an unbelievable attraction. The tower were completed by the famous railway-bridge builders Heenan & Froude from Manchester who made a special study of the construction techniques used on the Eiffel Tower. Particular importance was placed on fireproofing and the main legs were encased in concrete to stop the tower from falling on to the town even if the building at its base were to be burnt down.

To build it, I'm pretty sure that they would have employed steam power. They would have had a vertical boiler and a steam winch near to where the Tower Ballroom is now. All these arches that I've got at the bottom of my drawing can't be seen now because they're actually inside the building round the base of the tower. They would definitely have had what, as a lad, I used to call a steel erectors' stick; that's a big lump of wood, like a telegraph pole, with pulleys attached to the bottom of it and the same at the top. They would have started off with this standing on the floor with four guy ropes to hold it in place. The main haulage rope went over a wheel at the top and came to whatever winch or lifting gear they had. They'd pull the first bit of the steelwork up and when it was all bolted they'd lift up the stick inside the ironwork or close to it. This would have taken lots of manpower: blokes stacking pulley blocks at the top to stop it falling as it was advancing upwards and blokes at bottom with hand wheels. I've put the winch a little bit away from the tower, but I did it in a rush and think that actually they would've pulled it up the middle. When we were filming we went up to the maintenance level, which is basically the bit on the top of the building that surrounds the base of the tower and up there you can get some idea of what it's all about. Basically it's four latticework towers all leaning inwards and braced together with big 3-inch, $3^1/_2$-inch diameter diagonal tie rods to stabilize the whole structure and they tell me in a 70 mph gale it only moves an inch at the top thanks to all that lovely riveting. It's a credit to those Victorian engineers who built it.

For Fred, the Victorian Age represented the peak of engineering achievement. When it was built Blackpool Tower must have seemed like one of the great wonders of the world. The same is true of the Forth Railway Bridge, but, he said, 'I wonder what the Victorians would have made of the slim, graceful-looking suspension bridge that was built next to it in the sixties.'

Fred's drawing shows a section of the steelworks being pulled up with what he called a steel erector's stick

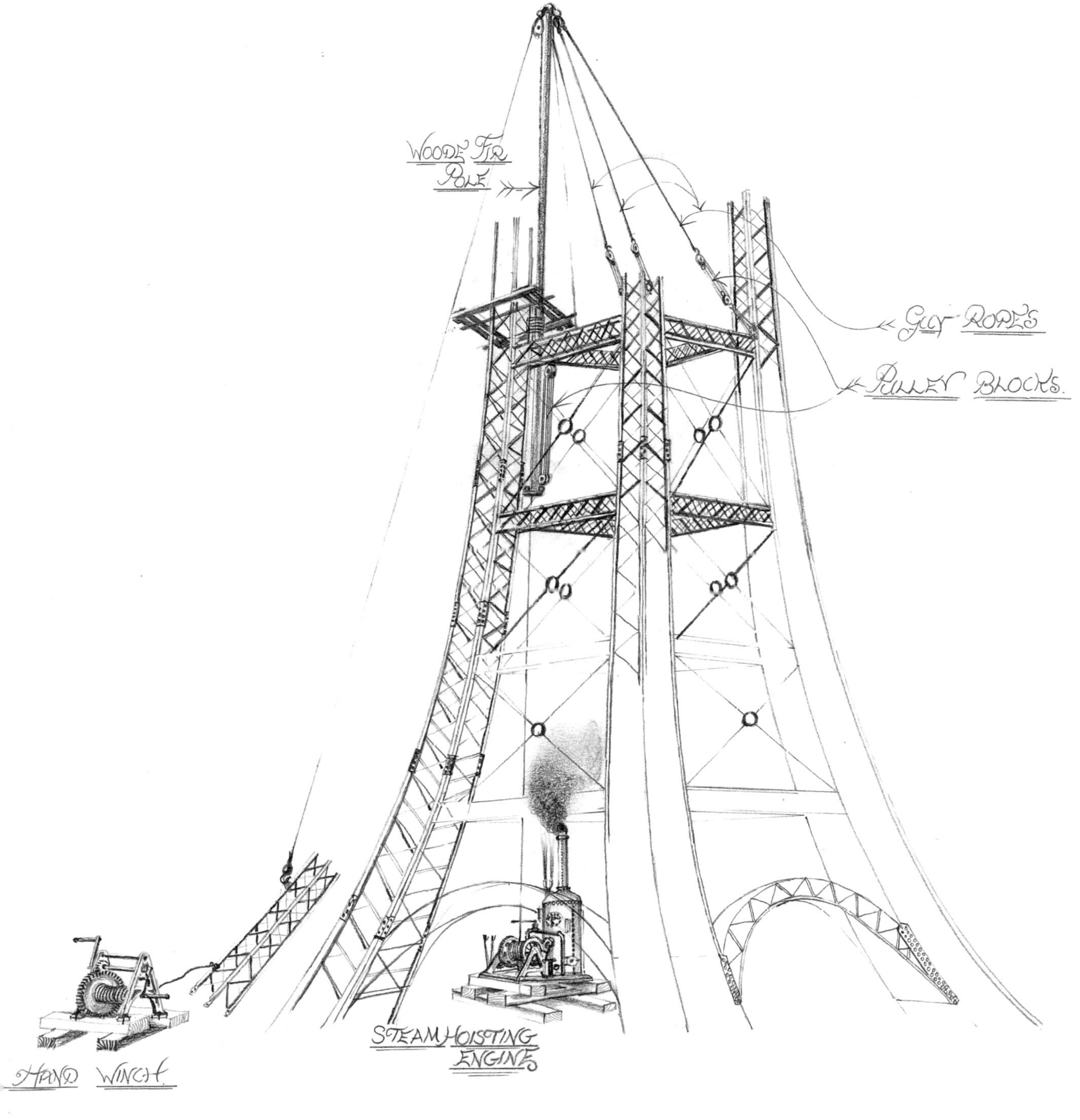

The Twentieth Century

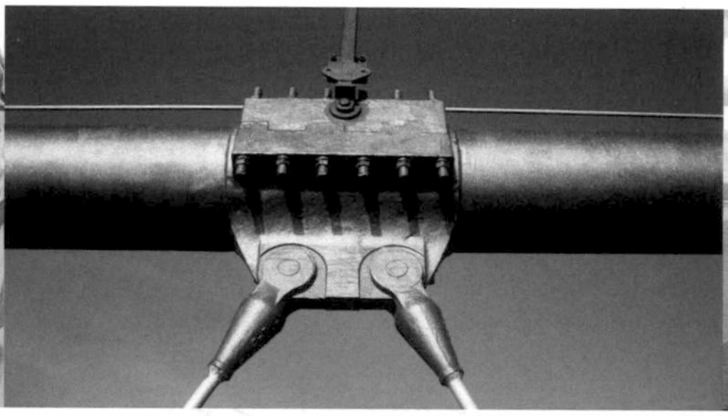

The Humber Bridge

On top of one of the towers of the Humber Bridge with bridge master, Roger Evans

The 1960s was a time when Britain led the world in the design and construction of suspension bridges. The biggest was the mighty bridge over the Humber which was the longest suspension bridge in the world. It has only recently been overtaken by one in Denmark and another in Japan. For a long time the Humber Estuary had been a barrier to trade between the two banks and local interests had campaigned for the construction of a bridge or tunnel across the estuary for over a hundred years. Approval for construction of a bridge was granted in 1959, although it was not until 1973 that the work finally began. Work proceeded for eight years and during this time many thousands of tons of steel and concrete were used and upwards of a thousand workers and staff were employed at times of peak activity. The bridge provides dual two-lane carriageways for highway traffic and there is a combined footpath and cycle-track along each side of it. The anchorages are massive concrete structures each containing two chambers within which the main cables splay out into separate strands. The piers are reinforced concrete structures which support the towers. Each of the towers consists of two tapered, vertical, reinforced concrete legs braced together with

Grace and strength – the 533-foot high north tower of the Humber Bridge

Fred's drawing shows how each of the bundles of wire that make up the cables fan out in the anchorages

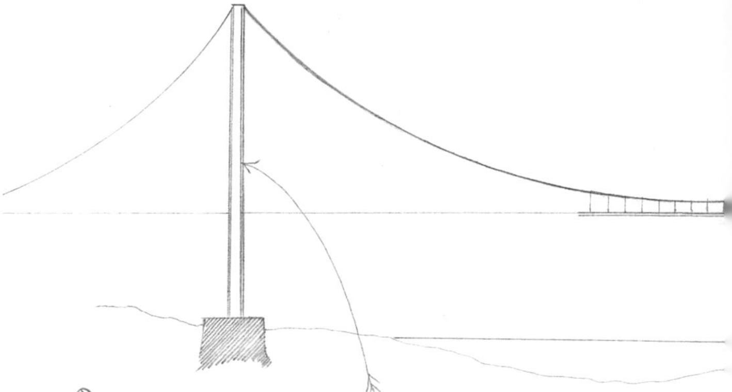

four reinforced concrete horizontal beams. Each of the main cables is made up of almost fifteen thousand parallel galvanized drawn wires, which for the purpose of erection and anchorage were divided into thirty-seven strands. The structure is suspended from the main cables by employing high-tensile steel wire strands and the suspended structure consists of stiffened steel plates welded together to form a hollow box section. Its streamlined shape makes the bridge aerodynamically stable and greatly reduces wind loads on it. Fred admired it:

> When you look at the slenderness and the gracefulness of the Humber Bridge it is quite fantastic to go down into the bowels of the earth at the end of it and see all those strands of wire that make up the cables that it is suspended from. They all fan out in this great underground chamber which has got dehumidifiers in it to stop any rust developing on the ends of them. The amazing thing about it is that it moves about 14 inches sideways in a gale, and goes back again. A very interesting piece of work. The towers are 533 feet high and I was taken up to the top by Roger Evans, the bridge master, and I asked him how they started with the first wire. He told me that it all started with a couple of men in a boat laying a wire rope on the bottom of the river, taking it over the top of the towers and pulling it tight. Then once they'd got several of them across they put an iron mesh on top of it to make a walkway, so that they'd got a catwalk across the river. That must have been a pretty hairy business. It's about a mile and a half between the two anchorages. Just think how much wire that is when you've got 14,000 wires forming a cable that is 2 feet in diameter going $1^1/_2$ miles. It's a lot of wire – enough to almost go twice round the world. The whole thing looks quite fragile but the cables alone weigh 11,000 tons, so the first thing they've got to do is support their own weight. And then you've got 17,000 tons of road hanging from them

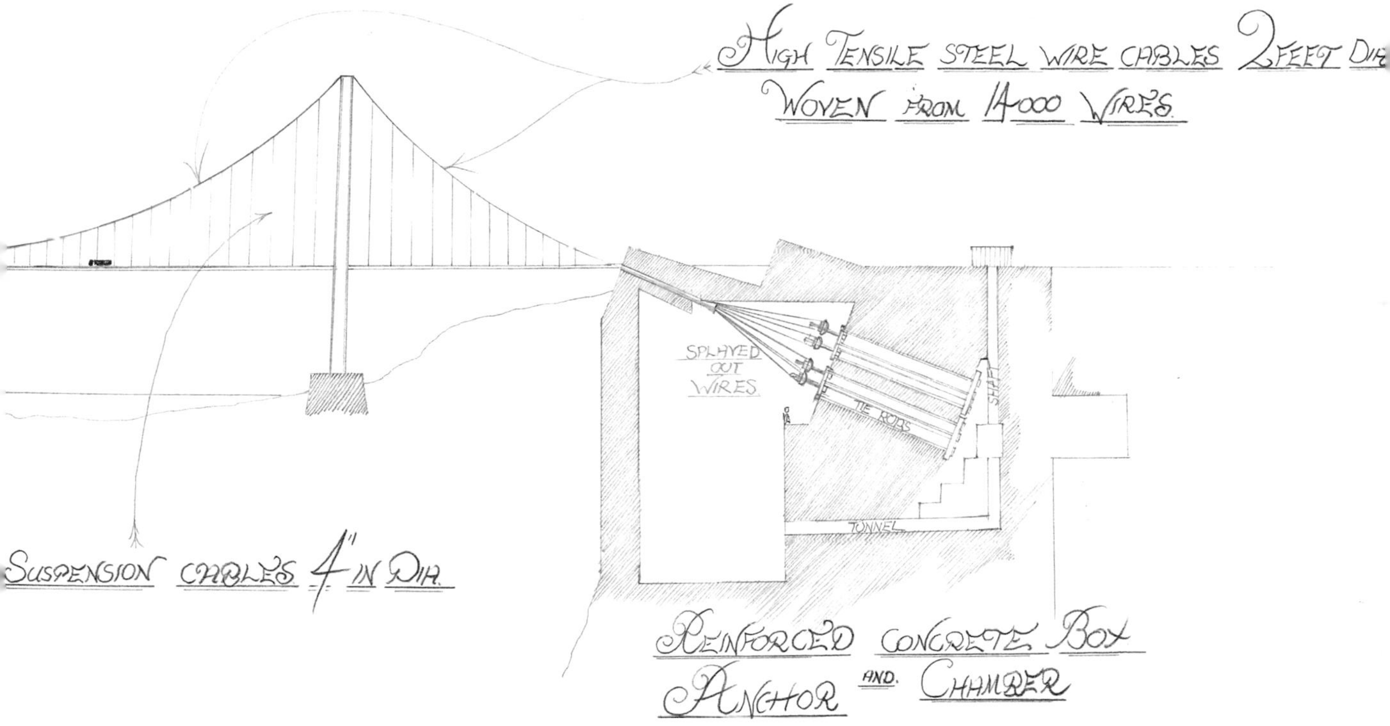

and on a good day there's 6,000 tons of traffic so we're talking big figures. The cable has unbelievable tensile strength – ten times what Telford's wrought-iron bars have got. It's not actually fourteen thousand separate pieces that it is made up from but one piece that traverses the whole width of the river from an anchor chamber on one bank, up, over, down and up and over and down and back again, two hundred times, making it four hundred passes in all. And each of these bundles of wire fan out in the great chambers at either end of the bridge – a great mass of wires like rays of sunlight coming from a funnel, up near the top. And it's from that slender-looking cable that the whole weight of the road is suspended.

The Thames Barrier

In this last thirty years or so we've seen some very impressive engineering achievements. One of the biggest projects has been the Thames Barrier. The great flood barrier across the Thames is one of the construction achievements of the twentieth century and a major advance in environmental control. The unique structure spans the 1,716-foot-wide Woolwich Reach and consists of ten separate moveable gates, each one pivoting on and supported between concrete piers and abutments which house the operating machinery. When raised, the four main gates, each weighing 3,700 tons, stand as high as a five-storey building and as wide as the opening of Tower Bridge. Four thousand men and women from all over

Protecting London – the Thames Barrier

Britain were engaged on the building of the barrier and the work, which cost nearly £500 million, took eight years. The construction of the reinforced concrete piers 50 feet below the waterline was the first stage of the process. To do this coffer dams, which are watertight boxes of interlocking steel plates, were first sunk into the bed of the river. The water was then pumped out and the piers were constructed.

The width of the river is divided by the piers to form four openings of 200 feet and two of 103 feet for shipping along with four subsidiary non-navigational openings. To the north of the river a huge dry dock was built in which the concrete sills were cast. After manufacture the dock was flooded and tugs towed the sills into position. They were then flooded and sunk to the level of the river bed 50 feet below. The piers and the sills formed the supports and seating for the gates and platform bases for the operating machinery so they had to be accurately built and positioned. The work had to be carried out in difficult conditions in all kinds of weather.

The gates' design is both simple and flexible in operation. To operate the barrage and close the gates, reversible hydraulic rams, one pulling and one pushing, are used to move rocker beams connected to discs at each end. These rotate the gates into any one of four different positions.

If Brunel had been around at the time it was being built, he'd have been involved in building it

Each gate has two sets of operating machinery, one at each end, either of them powerful enough for rotation.

Weather conditions are monitored twenty-four hours a day all the year round and in the control tower weather-forecast and tide-level information is received. A decision to close the barrier is taken by the duty controller and action will be taken about one hour after low tide, five hours before an incoming surge tide could reach this point. Fred described the operation:

> The Thames Barrier is a magnificent bit of engineering and in the Chief Engineer's office I saw a back-of-a-fag-packet drawing of the original conception. The drawing was more like something that Brunel would have done with beautiful towers to house the machinery on each of the piers rather than the present things which remind me of the hood on a windmill clad in stainless steel sheathing. Running right across the river under the gates there are tunnels that give access to the piers. The two tunnels are there to provide duplicate services. They may carry power and water for fire-fighting. So if one flooded, they could still close the barrier using the other one. On each of the piers there is a machinery room containing the hydraulic rams which are used to open and close the gates. When they are closed 3,600 tons comes rising up to close out the tide. They need to be closed when a spring tide and the strong easterly wind and low pressure all combine to create a higher than normal tide. When we filmed at the barrier in 1998 it had been closed thirty-three times to protect London from flooding. It can actually be closed in fifteen minutes, but that's a bit dangerous, because it can create a water-hammer effect on the Thames, so they like to take two hours to close it. There is a notch that can be seen around the piers and that's the height of the walls upriver towards London. The barrier's operators know if it's getting up there things are getting a bit dodgy. The width of the piers or the width of the gates are the same as Tower Bridge. It's the internationally known design, so if anyone builds a boat that is wider than this it won't fit through Tower Bridge and it won't fit through the Panama Canal either.

The Thames Barrier isn't really a bridge or a tunnel, but it's a great piece of engineering that combines elements of each.

'The whole thing,' said Fred, 'is a wonderful piece of civil engineering. One thing's for sure, though, is that if Brunel had been around at the time it was being built he'd have been involved in building it. It's just the sort of engineering challenge he would have wanted to take on.'

GAZETTEER

Here are the contact details for the places mentioned in the book, should you wish to visit or find out more.

AVEBURY
www.nationaltrust.org.uk
01672 539 250

BEAMISH OPEN AIR MUSEUM
www.beamish.org.uk
0191 370 4000

BEAUMARIS CASTLE
www.cadw.wales.gov.uk
01248 810 361

BLACK COUNTRY LIVING MUSEUM
www.bclm.co.uk
0121 557 9643

BLACKPOOL TOWER
www.theblackpooltower.co.uk
01253 622 242

BRIDGEWATER CANAL
www.bridgewatercanal.co.uk
www.penninewaterways.co.uk

CAERNARFON CASTLE
www.cadw.wales.gov.uk
01286 677 617

CHURCH OF ST WALBURGH
01772 726 370

CLIFFORD'S TOWERS, YORK
www.english-heritage.org.uk
01904 646 940

CONWY CASTLE
www.cadw.wales.gov.uk
01492 592 358

CONWY SUSPENSION BRIDGE
www.nationaltrust.org.uk
01492 573 282

CORNISH MINES AND ENGINES
www.nationaltrust.org.uk
01209 315 027

CRAGSIDE
www.nationaltrust.org.uk
01669 620 333

CROFTON PUMPING STATION
www.croftonbeamengines.org
01672 870 300

DOVER CASTLE
www.english-heritage.org.uk
01304 211 067

EASTNOR CASTLE
www.eastnorcastle.com
01531 633 160

EDINBURGH CASTLE
www.edinburghcastle.gov.uk
0131 225 9846

ELSECAR HERITAGE CENTRE
www.elsecar-heritage-centre.co.uk
01226 740 203

ELY CATHEDRAL
www.elycathedral.org
01353 667 735

ESCOMB CHURCH
www.escombsaxonchurch.com
01388 602 861

FORTH BRIDGE VISITOR CENTRE
www.forthbridges.org.uk

FOUNTAINS ABBEY
www.fountainsabbey.org.uk
01765 608 888

GLAMIS CASTLE
www.glamis-castle.co.uk
01307 840 393

HADRIAN'S WALL
www.hadrians-wall.org
01434 322 002

HAMPTON COURT
www.hrp.org.uk
0870 751 5175

HARLECH CASTLE
www.cadw.wales.gov.uk
01766 780 552

HARVINGTON HALL
www.harvingtonhall.com
01562 777 846

HEDINGHAM CASTLE
www.hedinghamcastle.co.uk
01787 460 261

HOUSE OF DUN
www.nts.org.uk
01674 810 264

HOUSESTEADS FORT
www.nationaltrust.org.uk
01434 344 363

HUMBER BRIDGE
www.humberbridge.co.uk
01482 647 161

IGHTHAM MOTE
www.nationaltrust.org.uk
01732 811 145

IRONBRIDGE GORGE MUSEUM
www.ironbridge.org.uk

KEDLESTON HALL
www.nationaltrust.org.uk
01332 842 191

LACOCK ABBEY
www.nationaltrust.org.uk
01249 730 459

LEVANT MINE AND BEAM ENGINE
www.nationaltrust.org.uk
01736 786 156

LITTLE MORETON HALL
www.nationaltrust.org.uk
01260 272 018

MARBLE CHURCH (WALES)
www.themarblechurch.org.uk

NATIONAL RAILWAY MUSEUM
www.nrm.org.uk
08448 153 139

NATIONAL WATERFRONT MUSEUM
www.museumwales.ac.uk
01792 638 950

NEWCOMEN ENGINE HOUSE
www.devonmuseums.net
01803 834 224

PETERBOROUGH CATHEDRAL
www.peterborough-cathedral.org.uk
01733 343 342

RICHMOND CASTLE
www.english-heritage.org.uk
01748 822 493

ST PAUL'S CATHEDRAL
www.stpauls.co.uk
020 7246 8350

THE SCIENCE MUSEUM
www.sciencemuseum.org.uk
0870 870 4868

SHILDON LOCOMOTION MUSEUM
www.locomotion.uk.com
01388 777 999

STOKESAY CASTLE
www.english-heritage.org.uk
01588 672 544

STONEHENGE
www.english-heritage.org.uk
0870 333 1181

THAMES BARRIER
www.environment-agency.gov.uk
020 8305 4188

TOWER BRIDGE
www.towerbridge.org.uk
0207 403 3761

TOWER OF LONDON
www.hrp.org.uk
0870 756 6060

WARWICK CASTLE
www.warwick-castle.co.uk
0870 442 2000

YORK MINSTER
www.yorkminster.org
01904 557 216

INDEX

ACKNOWLEDGEMENTS

I would like to thank the BBC for giving me permission to reproduce Fred's drawings. Fred's widow, Sheila, for her constant support and encouragement and for her own tireless efforts to ensure that Fred's memory is kept alive. My wife, Fran, for reading the manuscript, my daughter Kathryn for additional research and helping set up some of the photoshoots and our video editor, Pascal Morrison, for the frame grabs that were taken from the programmes. The team at Transworld, especially Doug Young for commissioning the book, my editor Emma Musgrave, copy editor Richenda Todd and designer Isobel Gillan, who has made the book look so good and has provided such a wonderful setting for both Fred's drawings and my text and photographs.

Also by David Hall

FRED
The Definitive Biography of Fred Dibnah

MANCHESTER'S FINEST
How the Munich Air Disaster broke the heart
of a great city

(written with Fred Dibnah)
FRED DIBNAH'S INDUSTRIAL AGE
FRED DIBNAH'S MAGNIFICENT MONUMENTS
FRED DIBNAH'S AGE OF STEAM